The Best of Kinks & Hints

The Best of Kinks & Hints

ALLAN B. COLOMBO

Butterworth-Heinemann
Boston Oxford Johannesburg Melbourne New Delhi Singapore

Copyright © 1998 by Butterworth-Heinemann

 A member of the Reed Elsevier group

All rights reserved.

No part of this publication may be reproduced, stored in a retrieval system, or transmitted in any form or by any means, electronic, mechanical, photocopying, recording, or otherwise, without the prior written permission of the publisher.

 Recognizing the importance of preserving what has been written, Butterworth-Heinemann prints its books on acid-free paper whenever possible.

 Butterworth-Heinemann supports the efforts of American Forests and the Global ReLeaf program in its campaign for the betterment of trees, forests, and our environment.

Library of Congress Cataloging-in-Publication Data
Colombo, Allan B.
  The best of kinks and hints / Allan B. Colombo.
    p.   cm.
  Includes index.
  ISBN 0-7506-9890-X (alk. paper)
  1. Buildings—Security measures—Miscellanea. 2.Fire alarms—
Miscellanea. I. Title.
TH9705.C64 1997
621.389′28—dc21   97-36247
                                                                       CIP

British Library Cataloguing-in-Publication Data
A catalogue record for this book is available from the British Library.

The publisher offers special discounts on bulk orders of this book.
For information, please contact:
Manager of Special Sales
Butterworth-Heinemann
225 Wildwood Avenue
Woburn, MA 01801-2041
Tel: 781-204-2500
Fax: 781-204-2620

For information on all Butterworth-Heinemann books available, contact our World Wide Web home page at: http://www.bh.com

10 9 8 7 6 5 4 3 2 1

Printed in the United States of America

Contents

Part 1	Access Control	1
1.	Access Control Installation	3
	When Not to Integrate for Egress	3
	Installing Electric Locks	4
	Assure Egress with a Double Break Circuit	4
	Buzzing Quiet Strikes	6
2.	Servicing Access Control	9
	Fixing Phantom Door Strike Releases	9
	Prevent Magnetic Stripe Card Misreads	10
	Magnetic-Stripe Card Care	10
	Clean up Access Control Equipment	11
	Dealer Looks to Solve Egress Motion Detector Problems	12
	Calculating Transformer Size	14
	Halting Relay Sparking	14
	Ventilate Transformer Enclosures	15
	Increasing Battery Power	17
	Meter Your Bad Fuses	17

Part 2 Burglar Alarms — 21

3. Burglar Alarm Installation — 23

- Put EOLs in Your Alarm Control Panels — 23
- Wiring In-Panel EOLs — 25
- Multiple-Switch EOLs — 26
- Dual EOLs for Better Line Supervision — 27
- Relay Solves Old EOL Problem — 29
- Supervising Hold-Ups — 29
- Assure Panel Integrity with Twisted Pairs — 31
- Tin Your Stranded Wires — 32
- Understanding Switch Nomenclature — 34
- Double Switch Your Doors — 34
- Protect Garages Without a Delay — 34
- Shunting Garage Doors — 38
- Maximize Your Photobeam Wires — 39
- Prolong Transmitter Life — 40
- Battery-Backup PBX Systems — 42
- Alarm-Controlled Guard Dogs — 42
- A Question of Sound — 43
- Ailing Power Causes False Alarms — 44
- Provide Arm-Only Capability — 45
- Properly Zone and Record Your Alarms — 46

4. Servicing Burglar Alarms — 49

- Prevent ac-Induced False Alarms — 49
- Suppress EMI in Cables — 50
- Isolate Problem Loops from RFI — 51
- Shorting out Your RFI Problems — 52
- Radio Frequency Interference Sources — 54
- Troubleshoot Defective Switches — 55
- How to Troubleshoot Swingers — 55
- Prevent Foil-Tape False Alarms — 57
- Stop Baffling Microwave False Alarms — 59
- Solve Those Difficult False Alarm Problems — 59
- Fix Troublesome Transmitters — 61
- Bad Communicator Connection — 63

5. Central Station Monitor — 65

- Dial-out Delay Avoids False Dispatches — 65
- Programming Error Causes False Dispatches — 65
- Better Protection for RJ31X Jacks — 67

Contents   vii

6.	Do-It-Yourself Circuits	69
	Protect Your Reed Switches	69
	Eliminate Bell Noise	70
	Put Capacitors to Work	71
	Switch Plate Foolery	73
	Add Voltage Regulation to Your Alarms	75
	Match Your Speakers and Drivers	77
	Switch Big Loads	78
	Control High Voltages with LED Power	80
	Build the Siren-to-Relay Circuit	81
	Build an Equipment Load Tester	82
	Build an Alarm Sensor Tester That Works	82
Part 3	CCTV	85
7.	CCTV Kinks & Hints	87
	Understanding Resolution	87
	Stop the Quad System Jitters	88
	Lightning Protection for Cameras	88
	How to Properly Set up Video Monitors	90
	Keep Space Between Coaxials and High-Voltage Wires	90
	Eliminate Harmful Ground Loops	92
	Resurrecting Tube Cameras	93
	Spare the Rod—Spoil the Camera	94
Part 4	Fire Alarms	97
8.	Fire Alarm Installation	99
	NFPA Clarifies "Family"	99
	Use EOL Markers	100
	Submit the Correct Fire Certificate	100
	Separate Your Feed and Return Circuits	102
	Smoke Detectors in Bedrooms	102
	How to Compute Strobe Intensity (1)	104
	How to Compute Strobe Intensity (2)	106
	ADA Clarification of "Adjoining Room"	107
	Here's How to Comply with the ADA	108
	Be NICET Certified	111
	Who's in Authority?	112
	Be Fire Code Smart	112
	Installing Beam-Type Smoke Detectors	113
	Mount Beam Smoke Detectors on Ceilings	114

	When and Where to Install Duct Detectors	115
	Upgrade Old Fire Pulls with Single-Wire Hookups	116
	Back up Your Electric Smoke Detectors	118
	Check Fiber Links on Phone Lines	120
	New SIA Communication Standard	120
	Terminate Wires Properly	121
	Give Supervisory Signals Distinct Sounds	121
	Power Fire Panels Ahead of the Mains	122
	How to Connect 25/70V Speakers to a Fire Alarm	122
	Use the Right Stranded Conductor	124
	File Your Fire Alarm Systems	125
	Use PLC Technology to Comply with the ADA	125
	Protect Plug-in Transformers	127
9.	Servicing Fire Alarms	129
	Updating Old Fire Alarm Systems	129
	Solving Fire Alarm Ground Fault Problems	131
	Properly Test Your Smoke Detectors	131
	Easy Test for Smoke Detectors	133
	Find Ground Faults Using the Half-Circuit Rule	134
	Increasing Battery Power	135
	Index	139

Part 1

Access Control

The need to control access, whether it be in a commercial, industrial, or military application, is integral to both property protection and personal safety. As security dealers more and more enter the intricate and difficult field of access control, the need for information increases proportionally. *Kinks & Hints* traditionally has offered dealers this information based on particular needs unique to the application of technology as applied to the control of access and egress.

Egress is of special importance because of the mandates set by the various building codes as well as the National Fire Alarm Code, NFPA 72, and Life Safety Code, NFPA 101, developed by the National Fire Protection Association, Quincy, MA. Egress, as it applies to access control and fire alarm protection, therefore, is first on the list of special segments presented in this third collection, *The Best of Kinks & Hints*.

1
Access Control Installation

WHEN NOT TO INTEGRATE FOR EGRESS

Security dealers who install fire alarm systems in large buildings usually are required to integrate them with an access control system when both exist in the same facility. This procedure helps assure that the occupants of the building get out safely when a fire is detected.

A representative of Corporate Protection Services Inc., Toledo, OH, says that, until recently, this was a simple matter of connecting the output contacts of a general-alarm relay inside the fire alarm control panel to an egress input at the main access control unit. The dealer says that, since the introduction of electromagnetic locks, however, integration is not always permitted by the authority having jurisdiction (AHJ).

Up front, NFPA 72 calls for the pressurization of stair towers in high-rise settings. Because nothing other than invisible magnetism is holding these doors closed, the premature release of one of these locks could mean the loss of air pressure in a stair tower. The dealer says that, because of this possibility, AHJs increasingly are asking security companies not to integrate their fire alarm systems with the main access control unit.

In its place, the dealer says he installs UL-listed egress buttons, touch-sensitive panic hardware, and egress motion detectors at stair tower exit doors. In many cases he has no choice in the matter because the AHJ specifically asks for these things. Either way, this recent development in the fire industry will do no less than earn security dealers additional revenue.

INSTALLING ELECTRIC LOCKS

Electromagnetic locks have become a big seller in the access control installation business. To be effective, however, they must have sufficient holding power and be mounted on a door that is strong enough to withstand the same forces.

Electromagnetic locks are manufactured in a variety of holding strengths, usually between 300 and 3,000 lbs. Electromagnetic locks with less than 1,000 lbs of holding force should be used for traffic control only.

Perhaps more important is the door to which the electromagnetic lock is mounted. In many cases the door material is weaker than the electromagnetic lock. For example, a typical aluminum or wooden door will fail under as little as 600 to 1,000 lbs of force. Steel doors are required for applications involving forces in excess of 1,000 lbs.

When you mount an electromagnetic lock it should be placed at the top of the door, opposite the hinge. Mounting electromagnetic locks in this position avoids the extreme forces that occur when a door is shoved, kicked, or rammed. The force is less at the top of a door because it flexes when it's violently hit.

Before purchasing an electromagnetic lock, be sure you know how many pounds of holding force it exerts. Always use an electromagnetic lock with at least 1,000 lbs of holding force and mount it at the top of a door that can handle the same forces.

ASSURE EGRESS WITH A DOUBLE BREAK CIRCUIT

Have you installed access control systems in public buildings? If the exit timers fail, can people get out of the building in an emergency?

Today, continuous-duty electric locks, such as electromagnetics, bolts, and dc-operated strikes, rely entirely on electronic timers to

Access Control Installation

release them. When timers fail, exit doors will remain locked, preventing egress.

"I always recommend the double-break circuit," says Bob Cook, president of Securitron Magnalock Corp., Torrance, CA. "The advantage is that, if the timer fails, the door still can be egressed by holding the exit egress switch down."

When an SPDT egress switch (see Fig. 1–1) is activated, it momentarily disconnects the electric lock from the system's low-voltage power supply. This simultaneously releases the door and

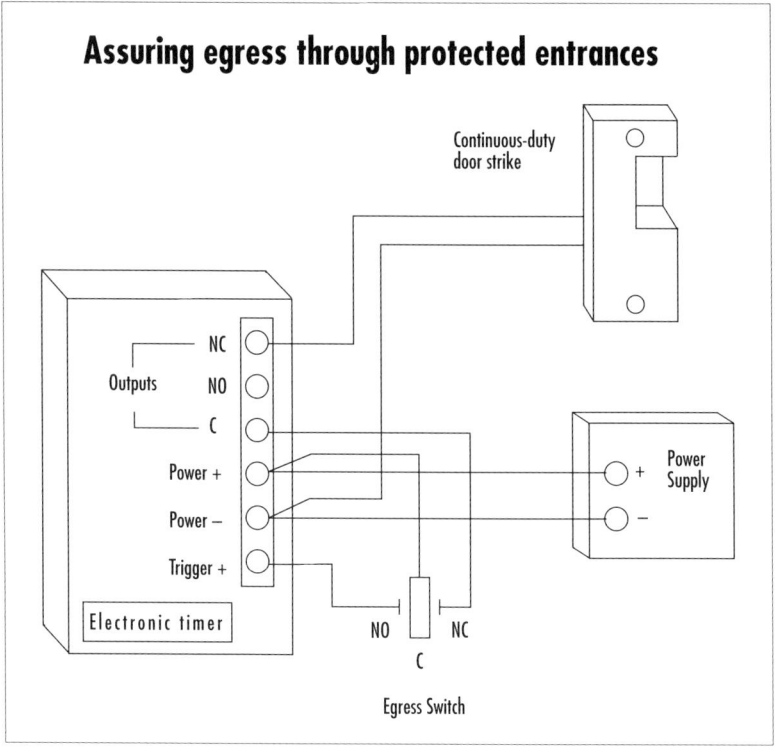

FIGURE 1–1. This circuit, presented by Bob Cook, President of Securitron, Magnalock Corp., Torrance, CA, assures egress, even when exit timers fail to operate.

triggers the timer. The contacts on the on-board timer relay then invert, severing the power connection to the door strike.

By the time the egress switch is returned to its normal, standby state, the timer has disconnected power from the strike. If, for any reason, the electronic timer fails, the egress switch itself will unlock a door.

BUZZING QUIET STRIKES

Electrically controlled exits "shall not require the use of a key, tool, special knowledge, or effort for operation from the inside of the building" (NFPA Life Safety Code 101, National Fire Protection Association, Quincy, MA).

For electric locks to comply with NFPA 101, they must be dc-operated "fail-safe" mechanisms. These continuous-duty devices typically release doors whenever power is lost or a mechanism fails.

Unlike ac-operated mechanisms, which buzz when a door is unlocked, most dc-operated locks operate quietly. This can be a nuisance to people waiting to pass through an entrance door.

One remedy is to install a separate buzzer next to an entrance door (see Fig. 1–2). The remote button might be an egress switch, motion detector, door-mounted touch-bar, heavy-duty relay, or the NC contacts found in an access control system controller.

When the remote button (switch) is activated, the door strike and relay are de-energized. The relay's SPDT contacts then switch electrical states, activating the buzzer until the strike and relay are energized again.

Access Control Installation

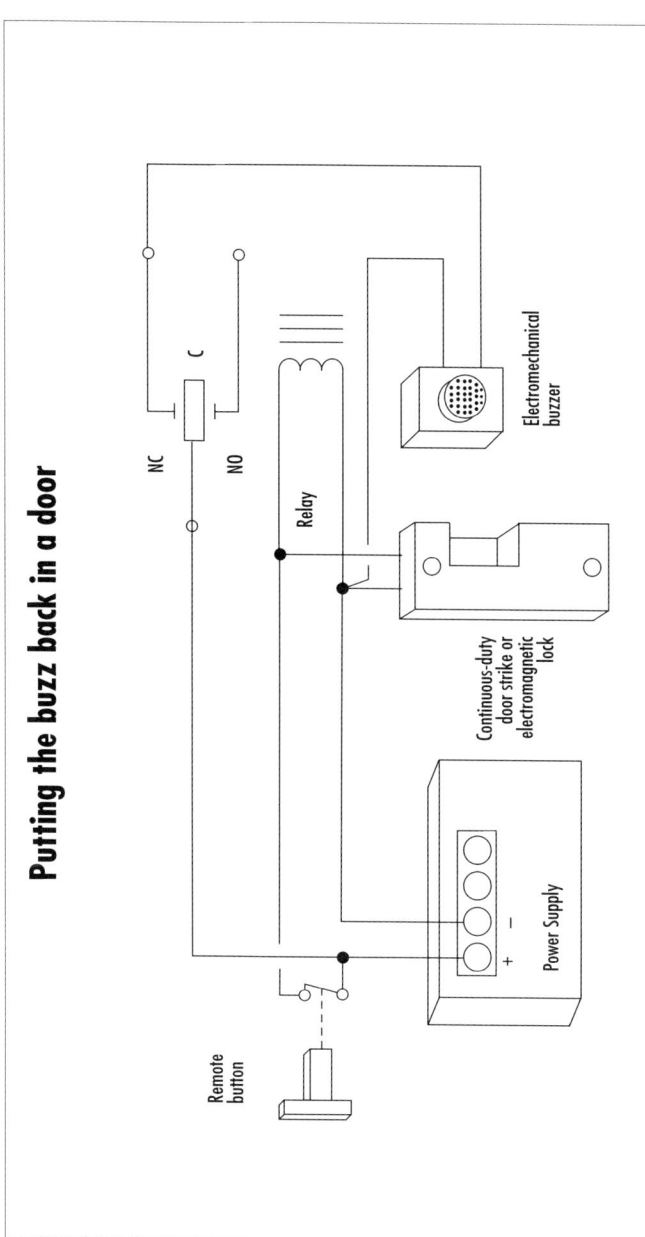

FIGURE 1-2. Adding an SPDT relay and dc buzzer gives an audible signal to doors equipped with continuous-duty, quiet-operating electric locks.

2
Servicing Access Control

FIXING PHANTOM DOOR STRIKE RELEASES

Vintage apartment access control systems can develop phantom-tripping after years of service. The problem may be associated with dirt sifting down from inside the apartment walls. This dirt, which usually first penetrates the intercom station's switching network, eventually combines with airborne moisture, forming a conductive solution across the switch contacts. This conductive solution is what eventually triggers the electric door release.

Finding that "certain" apartment station can be exasperating for a technician. Start at the intercom control amplifier by disconnecting each of the riser's "door" input wires, one at a time, until the offending riser is isolated. Then, reconnect the wire and locate the apartments serviced by this wire.

One at a time, remove the door wires of each apartment station until the bad intercom station is found. Then, either replace the door release switch or the entire intercom unit.

PREVENT MAGNETIC STRIPE CARD MISREADS

Security dealers should not shortcut their access control systems by buying inexpensive cards. They also should buy from well-established vendors with a proven track record. This is because magnetic stripe cards are vulnerable to a variety of problems that can affect the way data is encoded and then retrieved by access control read heads.

For example, the magnetic stripe on an inexpensive card may not be totally erased when it is re-encoded. This can happen even when erasing a card with a direct-current (dc) magnetic field. This is important because the stripe on these cards must be totally erased so the new information can be properly written to the same magnetic material (stripe). The quality of a card's lamination also is important because a card must remain flat when it's read by a reader.

Either of these problems can cause inconsistencies in the duration and amplitude (jitter) of the newly encoded data on a card. This can change how readers interpret the binary "ones" and "zeros," causing misreads. High-quality magnetic stripe cards, on the other hand, allow total erasure, and they use a high-quality lamination. This assures that the duration and amplitude of the data on a card remains intact and readable.

For more information, contact Q-Card Inc., 2 Chellis Ct., Owings Mills, MD 21117, or call (410) 581-0112.

MAGNETIC-STRIPE CARD CARE

Security dealers rarely hear about the mag-stripe cards they sell with their access control systems—until misreads start to occur. Card failures and misreads usually come from scratches in the surface of the magnetic tape.

"Light or moderate scratches should not cause misreads. [However], scratches through the stripe or deep into the stripe will cause an amplitude dropout for one or more bits," says a representative of Q-Card Inc., Owings Mills, MD. "This will cause a spot of high jitter and is likely to cause a misread."

Dealers can remedy this problem and add to their fiscal bottom line by offering protective card sleeves to their new and existing customers. "The most popular card sleeves that we sell our customers is made of paper or tyvek," says a company spokesman from CleanTeam Co., Simi Valley, CA. "Where the paper card is abrasive, however, the tyvek card is not."

Body oils and dirt on the surface of a magnetic stripe also can affect the integrity of the reads. Q-Card says that surface contamination is enough to cause misreads, rendering a magnetic stripe card unusable.

One way to protect mag-stripe cards from contamination, says CleanTeam, is to use a card sleeve with a card cleaner built right into it. "This is called passive cleaning because it does not require anything other than the same action that would be used to normally protect a card."

For more information on magnetic-stripe card care contact Q-Card Inc., Owings Mills, MD 21117, or call (410) 581-0112; and, for more information on card sleeves, contact CleanTeam Co., 960 Enchanted Way, Bldg. 108, Simi Valley, CA 93065, or call (805) 581-1000.

CLEAN UP ACCESS CONTROL EQUIPMENT

Prevent misreads in magnetic card readers by periodically cleaning them. The job is made easy with small credit-card-sized cleaning cards, like those manufactured by Clean Team, Simi Vally, CA. These cards are treated with a cleaning solution that removes dust and dirt from the surface of contact-type readers. They come in sealed airtight packages, which prevents premature evaporation of the cleaning solution.

To clean a magnetic reader, simply swipe/insert a cleaning card as you would a conventional card. Then wait until the solution on the card has dried before swiping/inserting again.

The manufacturer recommends weekly cleaning of inside readers and daily cleaning of readers in high-traffic areas. Some reader manufacturers also specify that cleaning take place once for every 1,000 uses. They also provide a wipe cloth for printer cleaning heads.

For more information, contact Clean Team Co., 960 Enchanted Way, Bldg. 108, Simi Valley Ca. 93065, call (805) 581-1000, or fax (805) 581-1058.

DEALER LOOKS TO SOLVE EGRESS MOTION DETECTOR PROBLEMS

A security dealer asked, "Recently we started using passive infrared detectors (PIRs) instead of momentary, mushroom-type egress buttons to release our clients' electromagnetic (EM) locks. However, every so often one of them will fail and we have to replace it. Can you suggest a way to prolong the use of these PIRs?"

I would venture to say that the PIR motion detectors you are using are of the off-the-shelf, conventional variety, rather than the egress type. Egress motion detectors are specifically engineered to permit people to exit buildings where the doors are electronically controlled by an access control system.

The relay contacts inside an egress-type motion detector have a higher current rating (1.5 A to 2.0 A) than their conventional PIR counterparts (0.10 mA to 125 mA). Having a motion detector with a capable relay output rating is critical because a typical EM lock draws in the vicinity of 150 mA at 24 Vdc/300 mA at 12 Vdc. Some EM locks can draw even more current.

Using an egress motion detector also is important because most models come with a Form C relay. Thus, dealers can use the same egress motion detector for both fail-safe and fail-secure applications. This will save the smart dealer money because he or she has to stock only one detector instead of several models. By comparison, conventional models usually come with a Form A, normally closed output.

One way to solve this problem on your existing units is to add a releasing relay equipped with contacts capable of safely passing 20 percent or more current than is specified for the electric strike or EM lock (see Fig. 2–1).

There is another reason why you need to consider using egress-type motion detectors in this type of application. In some of the larger municipalities, the fire authority having jurisdiction (AHJ) has been known to fail a fire alarm contractor's installation because another dealer used an unapproved egress motion detector in the access control system.

Dealers should remember that the matter of egress is regulated by NFPA 101, *Life Safety Code*, Chapter 5. So important is this issue to fire inspectors that many of them require that dealers use only approved egress motion detectors.

Servicing Access Control 13

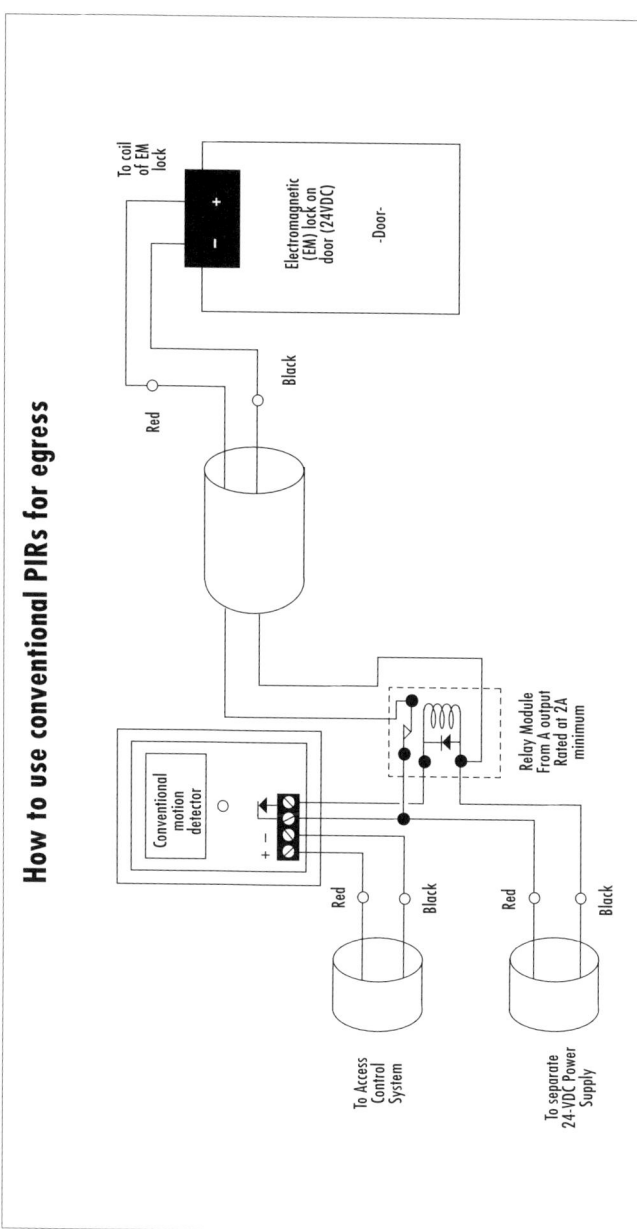

FIGURE 2–1. If a nonapproved egress motion detector is not an issue with the authority having jurisdiction, dealers can use conventional PIRs by adding a releasing relay equipped with a Form A or C relay with contacts that are rated for at least 2 A (first check with the AHJ).

"Where specifically permitted by Chapters 8 through 30, doors in the means of egress shall be permitted to be equipped with an approved entrance and egress access control system..." (Section 5-2.1.6.2, NFPA 101, *Life Safety Code*, 1994 Edition, Chapters 8 through 30, address specific applications and environments).

For more information, contact your local AHJ as well as the National Fire Protection Association, 1 Batterymarch Park, Quincy, MA 02269, call (617) 770-3000, fax (617) 770-0700, or visit its web site at http://www.wpi.edu/-fpe/nfpa.html.

CALCULATING TRANSFORMER SIZE

The plug-in power transformer used in most security applications is actually a "step-down transformer." These devices take 120 Vac and step it down to a lower voltage. Transformer size, which is measured in volt-amperes (VA), is an important factor when dealers install low-voltage systems and devices, such as ac-operated door strikes.

For example, if a 24-Vac door strike requires 0.2 A, the power (watts) required can be computed by multiplying voltage x current. In this case, 24 x 0.2 = 4.8 W. If two identical strikes are powered by the same transformer, the power consumption of the first one is multiplied by 2 (4.8 x 2 = 9.6 W).

In general, transformers should be capable of delivering twice the power of the circuits they power. This adds a margin of safety. In this case, 9.6 x 2 = 19.2 W.

Oddly enough, VA ratings are computed in the same manner as power consumption, (volts x amperes). That means that the VA rating of a transformer is virtually the same numeric value as the power consumption (W) of its load (19.2 VA).

In practice, a transformer's VA rating should be higher, never lower, than the power consumption of the device(s) it powers. Having more power available never hurts, but using a transformer with less VA than needed can cause a fire.

HALTING RELAY SPARKING

Relay contacts can be a problem when they are used to control large loads, such as those required for outside lights and large motors.

Sparking contacts also can result when large inductive loads produce back electromotive forces. This can etch and damage contact surfaces, eventually burning or welding the contacts together. A newly developed approach uses high-power semiconductor switches, in conjunction with conventional relay contacts, to eliminate contact sparking.

In the past, the remedy has been to install capacitors, diodes, metal oxide varistors (MOVs), or other transient suppressors over the contacts. However, installing the right capacitor, diode, or MOV can be a problem. The easiest solution is to use a sparkless relay.

A sparkless relay consists of a coil, contacts, and a high-power semiconductor device that switches the load "on" before the relay's contacts connect. Once the relay has taken over the current load, the solid-state switching device drops out.

VENTILATE TRANSFORMER ENCLOSURES

Nick Markowitz, president of Markowitz Electric Protection, Verona, PA, says to ventilate metal enclosures when you install plug-in power transformers in them ("Kinks & Hints," *SDM*, May 1994, p. 47). One way to do this is to drill and punch holes in the top of the box or on the cover or both. "Ventilation holes should be added to the cover over the transformer," says Markowitz.

Another way that Markowitz assures the integrity of the transformers he installs is to use locking thermostat covers (see Fig. 2–2). "I [also] use metal or plastic thermostat covers which are already punched and ventilated," says Markowitz. Furthermore, the lock prevents people from easily tampering with the wires or damaging the transformer inside.

One reason why ventilation is so important to conventional power transformers is their near-room-temperature design. For example, a conventional transformer installed in an unventilated metal box in a room at near room temperature usually will not cause a problem. However, when the ambient temperature in that room begins to increase, as on a hot summer day, the voltage and current from the transformer often will decrease.

This happens because the heat generated by the transformer combined with the outside room temperature increases the impedance of the transformer windings. An increase in impedance means less current, more voltage drop, and so a decrease in voltage at the alarm

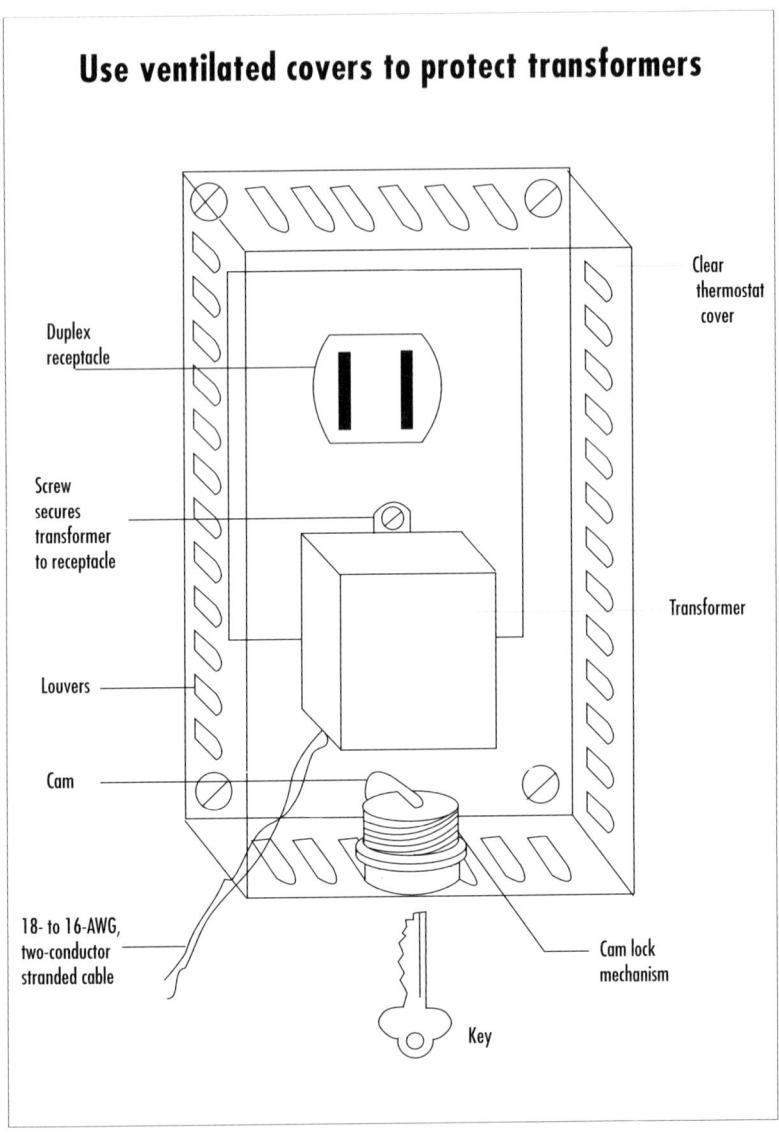

FIGURE 2–2. Nick Markowitz of Markowitz Electric Protection, Verona, PA, uses locking thermostat covers with louvers to protect plug-in transformers.

control panel. If the alarm control panel already is drawing current near the transformer's maximum rating, this can mean undercharged batteries and perhaps false alarms.

"A transformer is usually designed to have the highest efficiency at the power output for which it is rated. The losses in [a] transformer are relatively small at low output but increase as more power (current) is taken" (*The ARRL Handbook for Radio Amateurs, 1983,* Newington, CT, The American Radio Relay League, 1983 p. 2-25).

In extreme cases, the build up of heat is severe enough to cause the thin insulation on the coiled wire inside a transformer, or even the wire itself, to melt. "There is a limit to the temperature rise that can be tolerated because [a] too-high temperature either will melt the wire or cause the insulation to break down" (*The Radio Amateur's Handbook*).

In addition, Markowitz says that, to comply with the National Electric Code, NFPA 70, installers should ground the metal box that contains a transformer.

INCREASING BATTERY POWER

Increasing battery voltage and amp-hour ratings is essential when expanding battery backup systems. Fire alarm systems quite often operate on 24 V, requiring that two 12-V or four 6-V batteries be connected in series, for example. Battery polarity must alternate so the voltage of each battery adds to the next, equaling the sum of all the batteries.

The overall amp-hour rating of a battery backup system is increased by placing identical batteries in parallel. The amp-hour rating now increases while the voltage remains the same.

Both the amp-hour rating and the battery voltage can be increased by placing a number of batteries in series, increasing the overall voltage, and using an identical series-connected set of batteries in parallel, doubling the overall amp-hour rating of the battery supply.

METER YOUR BAD FUSES

Find bad fuses by measuring the amount of voltage over their circuit connections. This eliminates the need for removing good fuses while trying to locate a bad one. It also is helpful in low-light situations, when the threadlike filament in fuses cannot be seen so easily.

Use a digital volt/ohm meter (see Fig. 2–3) to measure the voltage over a bad fuse. Here's how to do it:

1. Switch the meter's function control to a voltage setting that exceeds the equipment's known source voltage (if the meter is not equipped with an autoranging feature).
2. Place the meter's probes over each fuse's circuit connections.

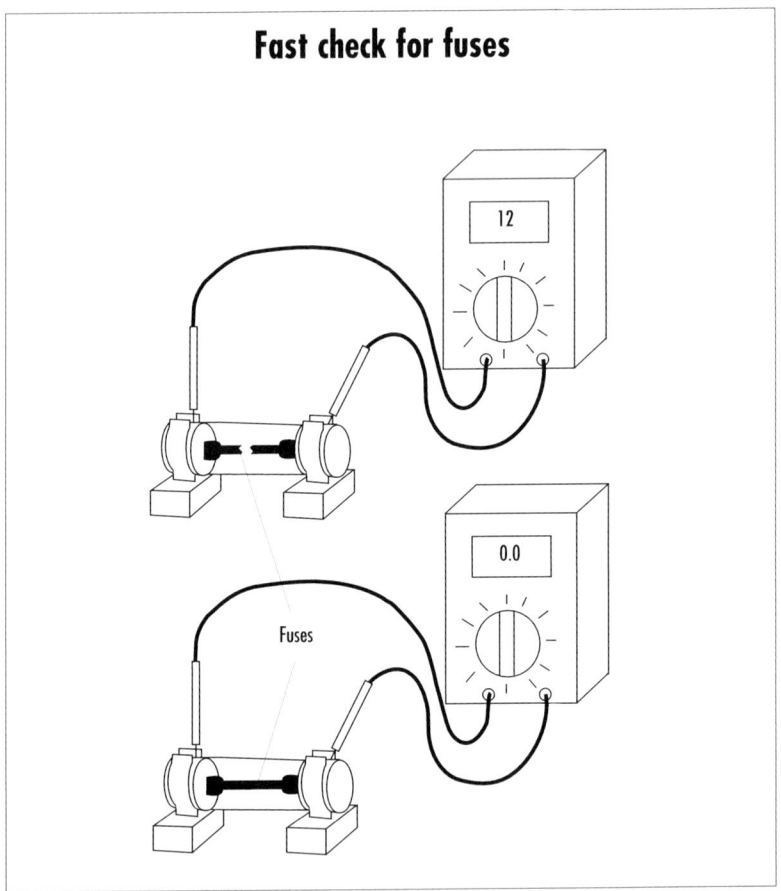

FIGURE 2–3. Find bad fuses by reading the voltage over connections. Source voltage indicates a bad fuse.

3. If the meter shows no (or very low) voltage, the fuse is operating normally.
4. If the meter shows source voltage, positive or negative, the fuse is bad.

Disposable glass fuses, for example, use threadlike resistive filaments that convert electron flow into heat. When the quantity of electrons flowing through them exceed a preset level, measured in amperes, the resistive element burns apart. This disconnects a load from its source.

When this occurs, the load no longer can draw current from the power supply. For all practical purposes, the load becomes part of a nonconducting electrical circuit that connects one side of the burned-out fuse and meter probe to the power supply. The other side of the fuse and meter probe then connect directly to the power supply. Fuses that indicate source voltage are burned out.

Part 2

Burglar Alarms

The need for practical how-to information is critical in the burglar alarm segment of the security industry. Most alarm technicians in the trenches look for installation and service ideas that will make their job easier, safer, and faster. When technicians work faster, alarm companies are bound to realize a higher percentage of profit, not to mention that the quality of the work they do also will be better. More times than not, this will result in additional work via word of mouth.

3
Burglar Alarm Installation

PUT EOLS IN YOUR ALARM CONTROL PANELS

In the past, security dealers have installed end-of-line (EOL) resistors at the ends of burglar alarm initiating circuits. Although in my opinion this is still the best way to install commercial alarm systems, one security dealer from West Allis, WI, says that he installs these devices near to or inside the alarm control panels to eliminate the frustration and time spent looking for them. He says he installs a second terminal strip in an alarm control panel or a second box next to the panel. He then installs a jumper wire from the negative (common) terminal of each pair of initiating zones and a corresponding screw terminal on the second terminal strip (see Fig. 3–1). One wire of each sensor circuit is connected to the positive zone terminal on the motherboard, and the other sensor circuit wire is connected to the EOL on the second terminal strip. This Allis, WI, dealer says that this better organizes the EOLs for identification and provides a test point for each initiating circuit, which he uses when checking the integrity of each sensor circuit. He does the testing by removing one end of all the negative (common) wires, either on the motherboard

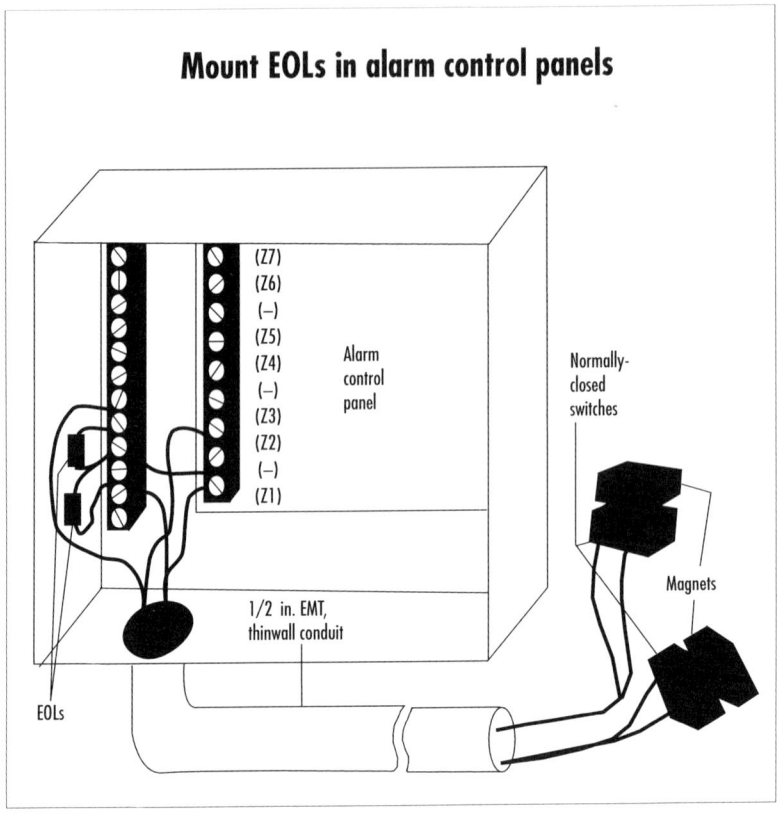

FIGURE 3–1. An Allis, WI, alarm dealer installs EOLs near to or inside the alarm control panels that he installs, which makes troubleshooting easier.

or the second terminal strip, and then checking the resistance of each sensor circuit by putting one end of a meter on the positive terminal (motherboard) and the other on the screw terminal where each circuit connects on the second terminal strip. Editor's Note: The method of connecting EOLs used by this Allis, WI, security dealer is highly controversial. Read on for additional tips on where and how to connect your EOLs.

Burglar Alarm Installation

WIRING IN-PANEL EOLS

An alarm installer recently asked, "In your article, 'Put EOLs Inside Your Panels,' Kinks & Hints, Nov. 1994, p. 43, you say that it's all right to install end-of-line resistors inside an alarm control panel. If this is so, then why do many panel manufacturers say to install EOLs at the end of a loop?" I doubt that anything I've ever said or allowed to be said in Kinks & Hints has ever created the fervor of "Put EOLs Inside Your Panels." I received letters from installers on both sides of the issue. For example, one dealer from AACC Security Systems, Nashville, TN, says "to install [your] EOL's in the panel, [run] the circuit out to the switch [using a four-conductor cable] and return it back to the panel (see Fig. 3–2)." This dealer's method, which he refers to as the "open/cross," requires the use of

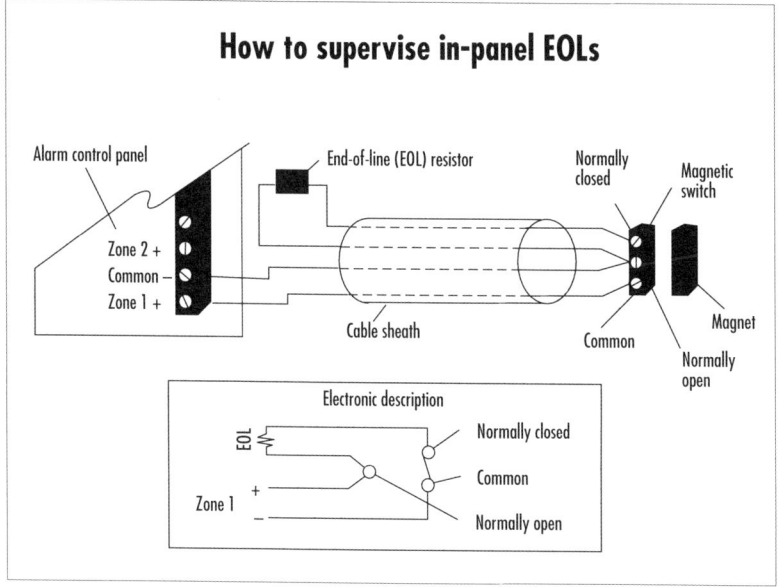

FIGURE 3–2. An alarm technician from AACC Security Systems illustrates how to install an EOL in an alarm panel. (Note: "Normally" is with magnet applied or relay power applied).

Single-Pole–Double-Throw (SPDT) contact switches. As a matter of policy, however, this dealer's company installs EOLs at the end of the circuit. "I agree with the Allis, WI, dealer, but the method you show does not provide end-of-line supervision for cable integrity," says another dealer from Alarm Services, Springfield, MO. "What happened to the high-side/low-side concept ("Make Proper Use of End-of- Line Resistors," Kinks & Hints, June 1993, p. 41)?" The Springfield dealer also says to install four-conductor cable, but in his method, current flows from the high side of the loop through one or more conventional SPST contacts and back to the panel through conductor #2, where it connects to one side (high side) of the EOL. Conductor #3 connects to the low side of the EOL, returning the circuit to the last device in the circuit. Here, it connects to conductor #4, which returns the circuit to the alarm control panel. Another way to track EOL locations is to keep detailed records. If this is not possible, then mark each EOL location (device) with a small, self-sticking color dot. The dealer from Nashville says that he's seen some installers even mark locations with fingernail polish. In an industry where there's almost as many ways to perform a specific task as there are installers, I can't help but think that the issue of panel-installed EOLs hinges largely on the specific application and the apparent security risks.

MULTIPLE-SWITCH EOLS

An alarm installer recently said that he was taught to make his systems as secure as possible by installing EOLs inside the alarm control panels. He asked if installing EOLs inside the alarm control panel makes them more vulnerable. Whether we like it or not, more alarm technicians are installing EOLs inside alarm control panels than you would have ever guessed. "Since 1981, almost 100 percent of the alarm systems I've taken over or serviced for someone else has had the end-of-line resistors installed inside the alarm panel," says Nick Markowitz, president of Markowitz Electric Protection, Verona, PA. The fact of the matter is that, when an end-of-line resistor is installed improperly, whether it's at the end of an initiating circuit or inside an alarm control panel, the wires are vulnerable to attack. They're also vulnerable to nondetection when there's a

Burglar Alarm Installation 27

ground fault condition on the line. In Fig. 3–3, technicians from Illinois Alarm and Stone Technologies Corp., Austin, TX, illustrate still another way to wire single and multiple switches with the EOL inside the panel.

DUAL EOLS FOR BETTER LINE SUPERVISION

Burglary, fire, and combination control panels that use dual end-of-line resistors supervise initiating zones better. This is because a minute current always is flowing through the circuit—whether the sensors attached to it are open or closed (see Fig. 3–4). Although use of dual EOLs is not yet widespread among security installers, a few control panel manufacturers already have added this feature to their panels. With some models, dealers can individually program each initiating zone for single- or dual-EOL operation. Alarm control panels with dual-EOL capability also provide more information. For example, in single-EOL circuits (see Fig. 3–4) normally closed (NC)

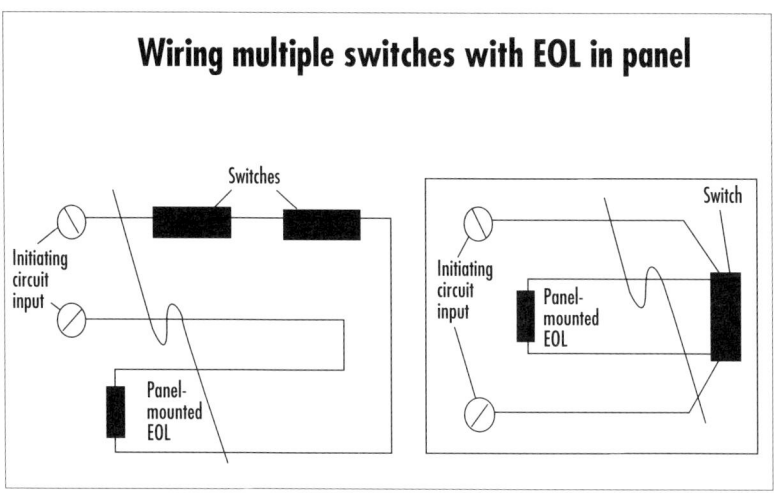

FIGURE 3–3. An Illinois alarm dealer illustrates how to wire multiple switches with the EOL in the panel (left) and an Austin, TX, manufacturer shows the single-switch method (right).

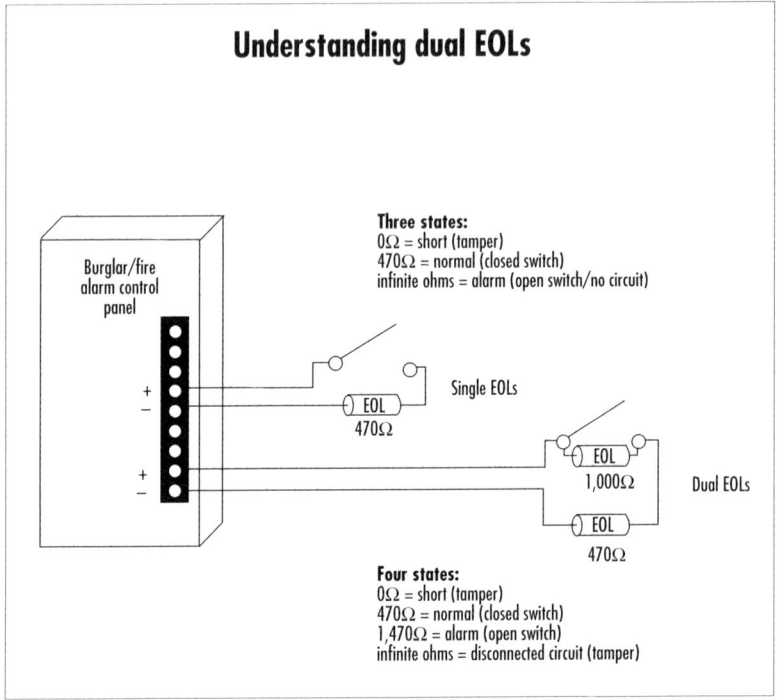

FIGURE 3–4. Because current in an initiating zone in a burglary/fire alarm panel equipped with dual EOLs (lower circuit) flows all of the time, a circuit can be protected against tampering, even when a NC switch is open and the alarm system is disarmed.

sensors commonly report up to three conditions: when switch contacts are closed, when switch contacts are open, and when there's a short between an initiating zone's EOL and its input. By contrast, dual-EOL-equipped initiating zones (see Fig. 3–4) can provide up to four circuit conditions: when switch contacts are closed, when switch contacts are open, when there's a short between an initiating zone's EOL and its input, and when the electrical connection between the input of an initiating zone and a contact switch/EOLs is severed. The added information that dual-EOL-equipped circuits provide enables service technicians to determine when someone has

Burglar Alarm Installation

tampered with an initiating zone. For example, a technically inclined criminal who sees the 470-Ω resistor (see Fig. 3–4, upper circuit) connected to one of the screw terminals of the door switch might try to place another 470-Ω resistor across the entire circuit between the alarm control panel and the sensor just before severing the circuit. If this happens during business hours when the system is disarmed, the control panel will not detect it. On the other hand, if the criminal sees the 1,000-Ω resistor (see Fig. 3–4, lower circuit) connected to the screw terminals of the door switch and installs another 1,000-Ω resistor across the entire circuit while cutting the line, the control panel will suddenly indicate a decrease in resistance on the line (1,470 Ω to 1,000 Ω).

RELAY SOLVES OLD EOL PROBLEM

An alarm dealer recently asked, "How can a new alarm panel be made to work with an old [original] EOL that's of the wrong resistance?" Nick Markowitz, of Markowitz Electric Protection, Verona, PA, found a way to make a replacement alarm control panel work with an old EOL that he could not remove from a door frame (see Fig. 3–5). Markowitz used a sensitive relay, model RBSN-TTL by Altronix Corp., Brooklyn, NY, and the new panel's 12-Vdc auxiliary power supply to make it work. Normally, passing a dc current through a switch is not an ideal situation, but because of the 4.7 K-Ω resistor and the low-current, sensitive relay that Markowitz used, the current in this circuit is less than 2.5 mA. Contact Altronix Corp., 140 58th St., Bldg. A3 West, Brooklyn, NY 11220, call (718) 567-8181.

SUPERVISING HOLD-UPS

When installing silent hold-up systems use a supervised loop from an alarm control panel, says Integra Controls, Austin, TX. The use of unsupervised inputs is not uncommon in older installations, where alarm installers often connected hold-up buttons directly to digital communicators and power supplies.

Today, there are better ways to install hold-up systems. Whether you use a radio receiver and wireless hold-up button, a hardwired latching hold-up button, a foot rail, or a money clip, the use of a

How to make the wrong EOL work

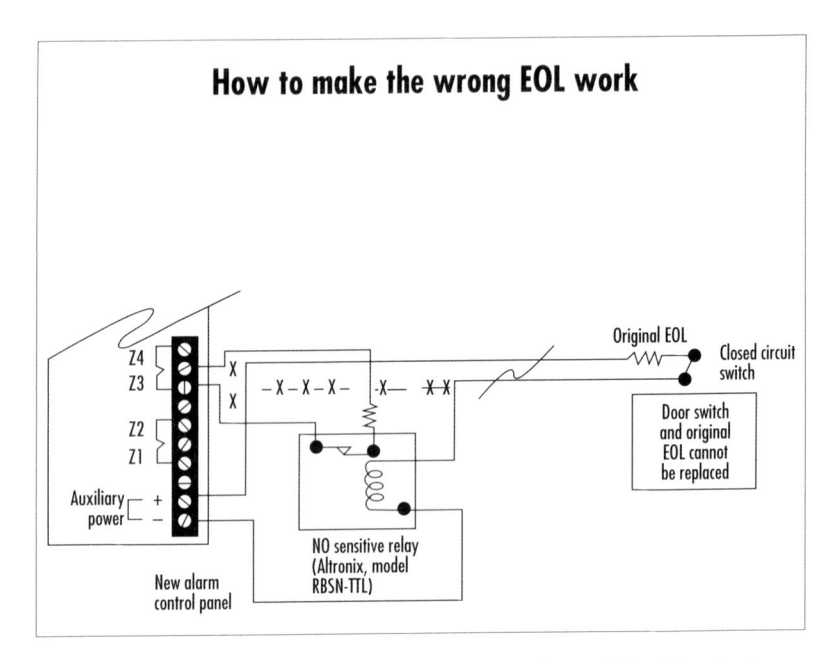

FIGURE 3–5. The RBSN-TTL sensitive relay, manufactured by Altronix Corp., Brooklyn, NY, enables Nick Markowitz, Verona, PA, to use an old EOL buried in a door frame.

supervised end-of-line resistor offers greater circuit integrity. No one can guarantee that a hold-up switch will work when it's pressed. But a supervised circuit will improve the reliability of the electrical circuit between the control device and the switch. Improvements in hold-up buttons, such as sealed magnetic reed switches, have helped to assure that a hold-up button will work when it's needed. A microprocessor-based alarm panel also provides an automatic test report to the central station on a regular basis. Unlike a simple digital communicator and power supply system, an alarm panel can supervise its own internal circuits as well as the telephone line itself (see Fig. 3–6). A problem in the system or telco line then is reported to the central station or management.

Burglar Alarm Installation 31

Alarm panel supervises hold-up buttons

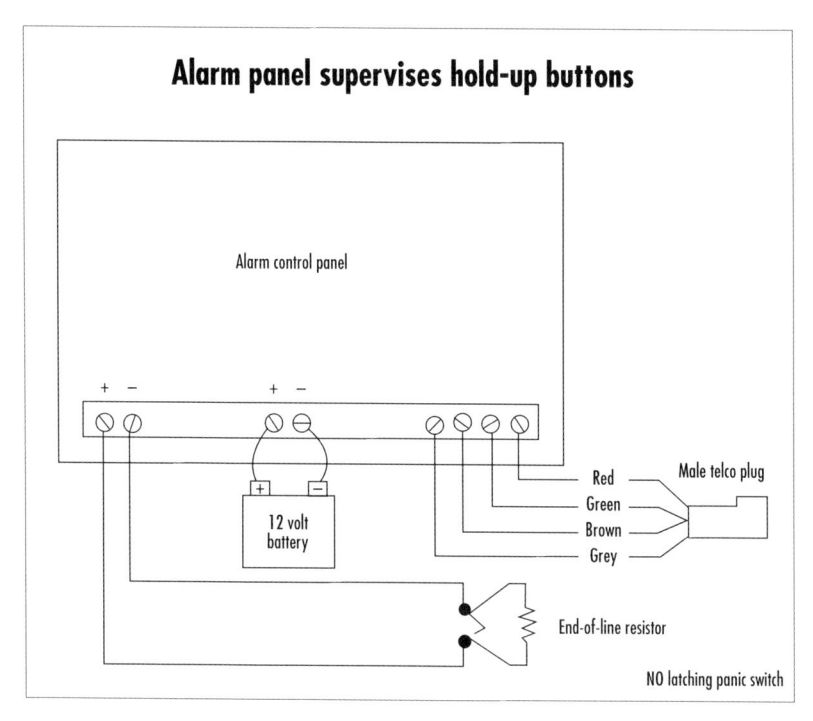

[Diagram: Alarm control panel connected to 12 volt battery, with Red, Green, Brown, Grey wires going to Male telco plug; End-of-line resistor and NO latching panic switch shown]

FIGURE 3–6. Using an alarm control panel in a hold-up system provides battery backup and the use of a supervised alarm circuit.

ASSURE PANEL INTEGRITY WITH TWISTED PAIRS

An alarm installer recently asked, "Can I use quad telephone station wire for my keypads, motion detectors, glass-break detectors, and various other burglar alarm devices that require four-conductor cable?" You sometimes can get away with installing four-conductor station wire, providing your wire runs are relatively short and you avoid electrical wires and other sources of electromagnetism (EM). Twisted-pair cables, however, will naturally reject EM interference better than quad cable. For example, the four wires in a quad station cable usually are laid parallel to one another in the cable sheath. On the other hand, each pair of wires in a twisted-pair cable are twisted together (see Fig. 3-7). The twisted-pair cable rejects EM interference better than straight station wire because the induced current in a twisted pair cancels out. Both wires effectively share the same center axis; so, because the current induced by the interference in each is opposite to that of the other, the interference-induced current is effectively canceled out. This canceling effect is especially important in communication equipment, keypad circuits, and other applications where low-level digital data or analog signals travel. For more information, contact Leviton Telcom, 2222 222nd St. S.E., Bothell, WA 98021, call (206) 486-2222, or fax (206) 483-5270.

TIN YOUR STRANDED WIRES

To prevent short circuits and false alarms, tin your stripped-out stranded wires before installing them on terminal strips in alarm control panels. The hairlike strands on these wires can short out to each other when they are installed on adjacent terminals (see Fig. 3-8).

Reducing EM interference

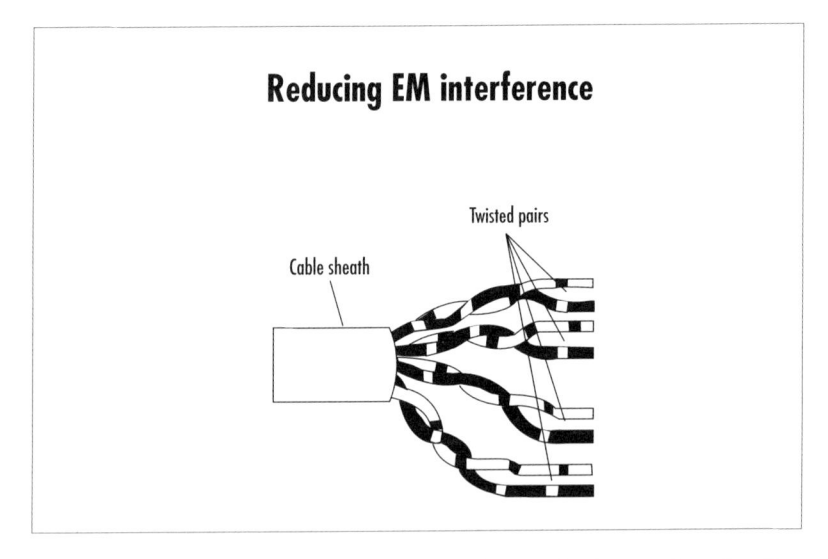

FIGURE 3-7. Use multiple-conductor cable with twisted pairs to reduce the effect of induced electromagnetic interference.

Burglar Alarm Installation

Tin wires to prevent shorts

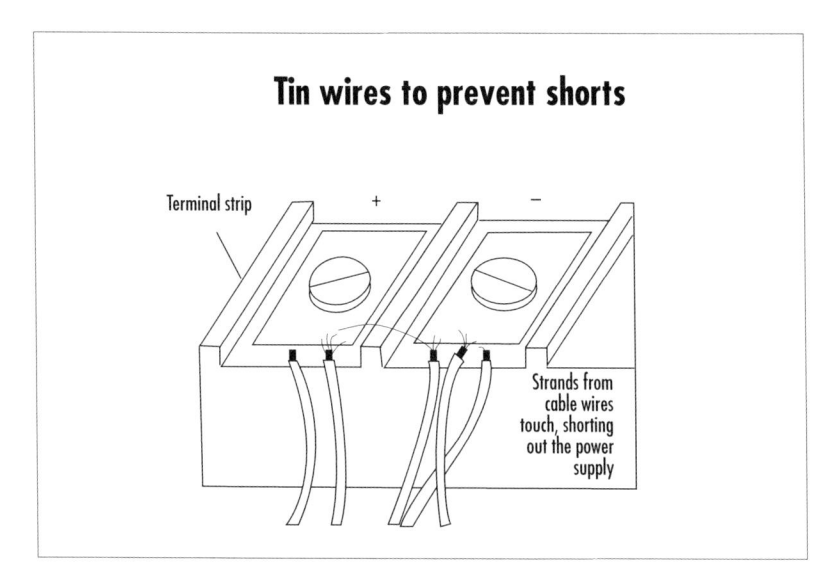

FIGURE 3–8. Stranded wires on adjacent terminals can short to one another when they are not tinned.

This can cause fuses to burn out when they're connected to auxiliary-power and bell-output terminals. Wire ends that are not tinned also can lead to false alarms, especially in alarm systems installed near industrial plants, large freeways, railroad tracks, and other facilities that generate seismic vibration. This usually becomes a problem when the hairlike strands begin to break, one at a time, as they are exposed to the constant back-and-forth motion of a vibrating structure. Eventually only one or two wire strands are left to carry the alarm-detection current, which can cause the resistance in the circuit to fluctuate, causing false alarms. To tin a stripped-out, stranded wire, first twist the strands together with your thumb and forefinger. Then heat and solder them together using a soldering iron and 60/40 (lead/tin) solder. After soldering, be sure no ragged ends remain. Then, square off the tip of the wire by cutting it with a diagonal cutter. This assures that the wire sits squarely against the back of the terminal strip under the metal plate.

UNDERSTANDING SWITCH NOMENCLATURE

Novice installers often are bewildered by "fail-safe" motion detectors that use energized output relays. This is because the electrical state of their common (C) moveable armatures and normally open (NO)/normally closed (NC) contacts, printed on circuit boards and terminal strips, may not agree with an installer's VOM. Fail-safe models, for example, are designed so the relay drops out, activating an alarm when a burglar cuts a low-voltage power wire. "Fail-secure" models, however, use de-energized relays that do not change electrical state when power is removed. The National Electrical Manufacturers Association (NEMA), Washington, DC (NEMA Standard ICS 1-1988, p. 24) instructs manufacturers to "show contacts by a symbol indicating the circuit condition when the device is in its de-energized or non-operated position." This usually means without power. The schematic symbols for form "C," "B," and "A" outputs (in their unviolated, detector-powered state) is illustrated in Fig. 3–9. Both "de-energized" (fail-secure) and "energized" (fail-safe) devices are shown.

DOUBLE SWITCH YOUR DOORS

When a break-in occurs, a criminal can easily end up on the wrong side of an interior door. This is a major problem when a magnetic surface switch is exposed to a criminal who knows enough to short the switch. Security dealers usually try to anticipate this problem by logically establishing which portion of a facility a criminal is most likely to break into. However, no one can really predict at which side of an interior door a criminal will end up. To prevent the worst scenario from occurring, install two magnetic surface switches on your interior doors (see Fig. 3–10). Now, if a criminal shunts the switch on the left side of the door, there's another one on the other side that he or she knows nothing about. Be sure to junction both switches in an attic or a secure junction box equipped with tamper-proof screws so a criminal cannot short both of them.

PROTECT GARAGES WITHOUT A DELAY

Here's a neat way to provide instant detection in a garage with a PIR without the typical long delays that dealers usually have to use.

Burglar Alarm Installation

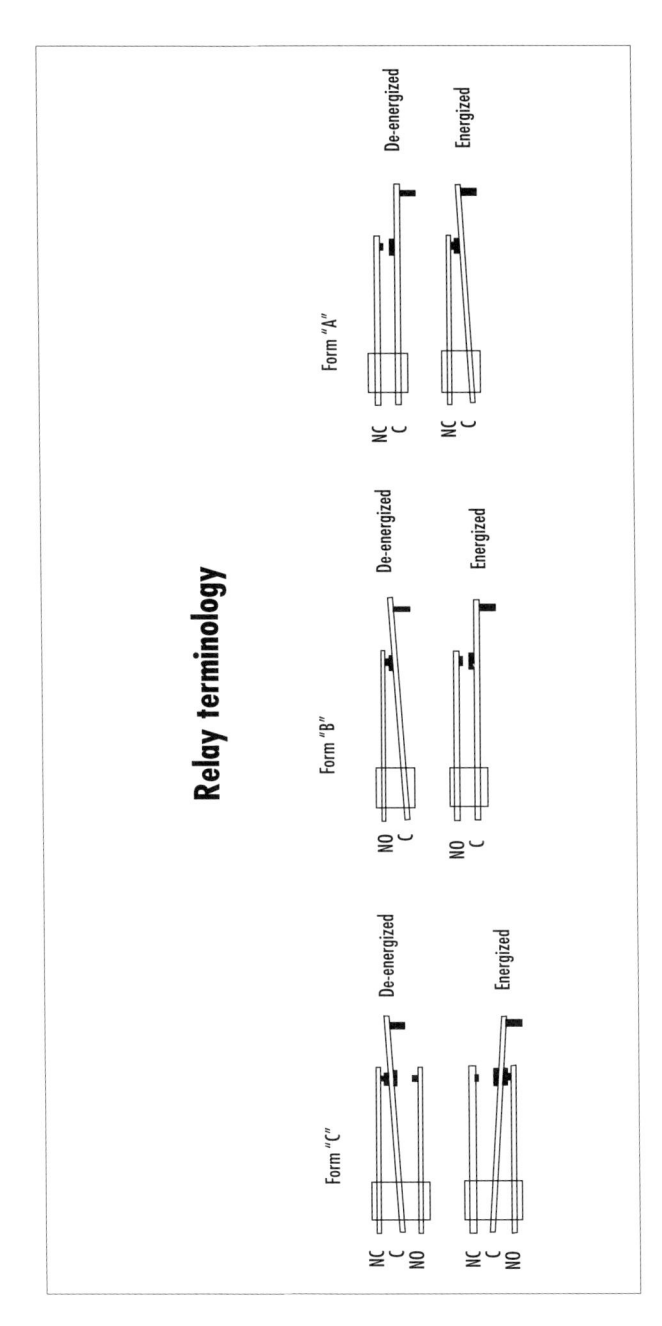

FIGURE 3–9. Here are the various types of contacts used in the energized and unenergized relays found in detection equipment in the security industry.

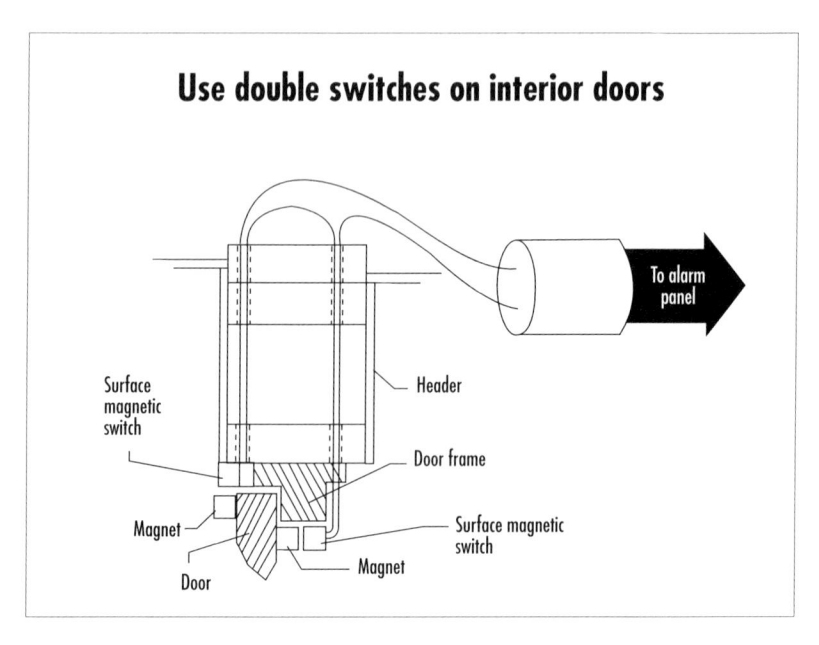

FIGURE 3–10.   Use double switches on interior doors.

"This is a circuit I use when protecting a customer's garage with a PIR. Now he can get in and out of his garage with ease without rushing to a keypad," says Nick Markowitz of Markowitz Electric Protection, Verona, PA. When the garage door begins to open, a wide-gap switch supplies power to a 12-Vdc relay (see Fig. 3–11). The normally open (NO) and common (C) contacts on the relay then shunt out the PIR's normally closed (NC) contacts. The LEDs on the switch plate in the house indicates the status of the garage door so the homeowner knows if it's open or closed. When the homeowner wants to leave, all he or she has to do is arm the system and press the momentary button on the switch plate to open the overhead door. This causes the garage door to travel upward, shunting the PIR. Because this occurs before the homeowner enters the garage, he or she can enter the garage and pull the car out without setting off the system. Then, when the garage door closes, the PIR is restored to the alarm circuit.

Burglar Alarm Installation

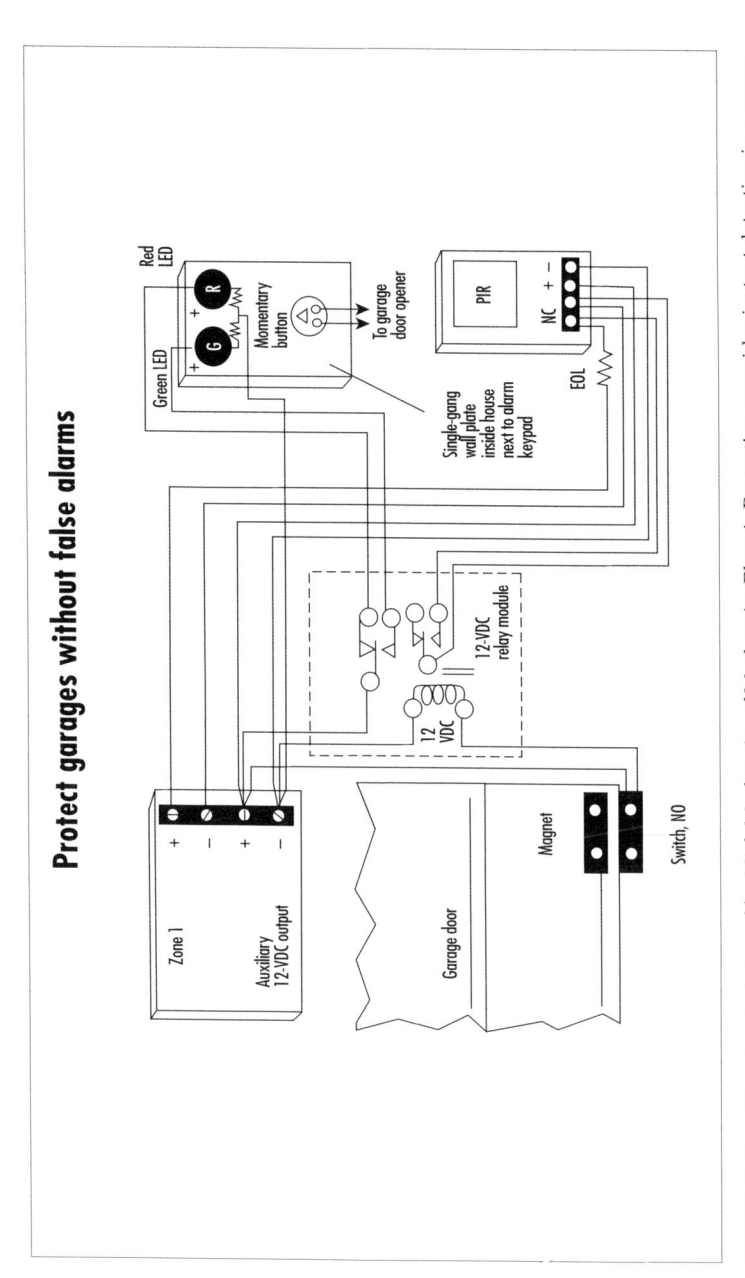

FIGURE 3-11. This circuit, designed by Nick Markowitz of Markowitz Electric Protection, provides instant detection in a garage using a PIR.

SHUNTING GARAGE DOORS

Installers usually do not protect garage doors equipped with motorized garage door openers that automatically open a garage door when a homeowner presses a transmitter button in his or her vehicle. This is due to the usually long entry delay before the homeowner can park a car and disarm the alarm system. A long entry delay also gives a criminal more time to enter and disarm or circumvent an alarm system. One way to protect a garage door without a long delay is to use an electronic timer, balanced door contact, low-voltage relay, voltage-dropping resistor (if the trigger voltage exceeds the relay's coil), and a garage door opener that emits a trip voltage output when activated (see Fig. 3–12). During standby, the floor-mounted magnetic switch is

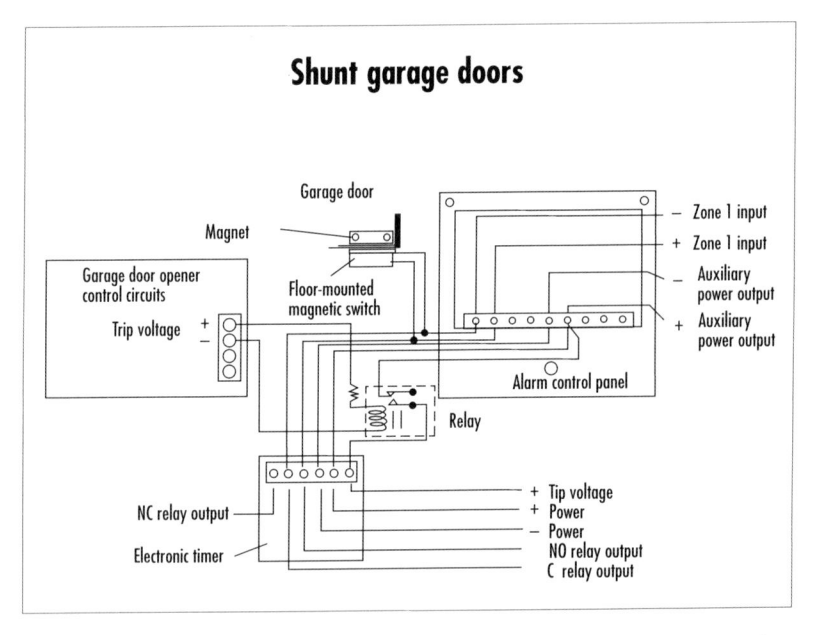

FIGURE 3–12. You can shunt garage door sensors with an electronic timer and a low-voltage relay.

monitored by the alarm control panel. When the garage door opener is activated, however, it sends a positive trip voltage to the timer, which triggers the timer's relay. The NO output from this relay then closes, shunting the switch on the garage door. This prevents the alarm control panel's initiating circuit (zone 1) from detecting the open garage door.

Because the garage door opener's trip voltage will disappear sometime after the opener stops running, the programmable timer continues to shunt the garage door switch for a predetermined period of time, allowing the shunt to remain for as long as the timer is set. This allows a homeowner to park a car and either disarm the alarm system or shut the garage door by activating the garage door opener again. Reactivating the garage door opener refreshes the timer's delay time, which continues the shunt. The trip voltage from most garage door openers usually is 12 to 24 Vac/Vdc. If the voltage is ac, for example, connect a diode in series with the relay's coil to convert the ac current to dc. Also, if the trip voltage exceeds the relay coil's specified voltage, use Ohm's law to adjust the value of the voltage-dropping resistor so it drops the difference between the coil voltage and the trip voltage.

MAXIMIZE YOUR PHOTOBEAM WIRES

An alarm installer recently asked, "The outdoor photoelectric beam detectors I use come with an environmental output to shunt the alarm relay when there is heavy fog, snow or rain. Although this reduces the likelihood of false alarms, I am concerned that there is no provision with which to notify the monitoring station when an environmental problem takes place. Do you know of a more secure way to do this?" One way to do this is to run a multiple-conductor cable, as shown in "Photobeams Are Hard to Beat for Outdoor Detection" (*SDM*, May 1996, p. 59), and install relays inside the alarm control panel. This provides multiple relay outputs, allowing the dealer to configure the environmental contacts to shunt the alarm contacts while tripping a 24-hour zone (no siren output) inside the alarm control panel. An alarm technician from Central Signaling Monitoring Co. (Cen Signal), Columbus, GA, says that he knows a better way to wire an outside photoelectric beam detector, with an environmental circuit, to an alarm control panel with fewer wires. In addition, his method will enable you to supervise the detection wires (see Fig. 3–13). The dealer's method uses the relays built into the photoelectric beam detector to avoid adding relays at the alarm control panel. He can do this because the negative side of each zone and the negative side of the auxiliary power supply are connected together, which allows him to reduce the number of wires from eight to five. This also enables dealers to install the panel's EOLs in the field inside the photoelectric beam detectors instead of inside the alarm control panel at the relays.

PROLONG TRANSMITTER LIFE

Wireless security transmitters commonly are installed in lieu of wires. These devices, however, are sophisticated subsystems all to

Maximize Use of Environmental Relays

```
Outdoor photoelectric
beam detector
Power                              + Power
supply                             − Power
           EOL
                                   Zone 1 alarm
Alarm
           EOL                     Zone 3
                                   Environmental
Environ-
mental                             Zone 2 Tamper
           EOL
Tamper
                                   Alarm control panel
                                   inputs and outputs
```

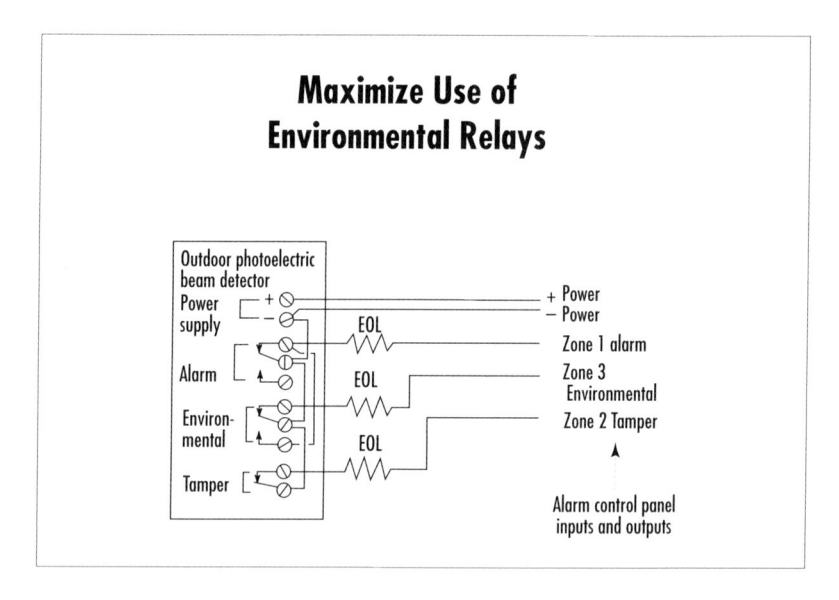

FIGURE 3–13. An alarm technician from Cen Signal, Columbus, GA, illustrates how to connect the relays in an outdoor photoelectric beam detector so the environmental relay shunts out the alarm zone and notifies the central station when there is an environmental problem.

Burglar Alarm Installation

themselves that can be damaged inadvertently by placing them in a damp environment, where frost and condensation can form around or inside them. Transmitter screw and solder connections, for example, can become corroded, and harmful deposits can form on circuit boards. When moisture presents a potential problem, install plastic, nylon, or some other moistureproof washers to keep transmitters off of door and window frames. Dealers also can partially protect security transmitters by sealing the circuit board, and terminal connections within them, with a polyurethane spray. Here's how to do it:

- Discharge the static electricity from your hands before you begin. This will protect the sensitive components on the printed-circuit board inside the transmitter.
- Take off the transmitter cover and remove the battery from the transmitter.
- Observe polarity if it's a lithium battery.
- Gently remove the circuit board from inside the transmitter housing.
- Use clear plastic or masking tape ($\frac{1}{2}$ in. wide) and mask the terminal screws, programming-plug contacts, battery connections, and tamper switches. This procedure will depend on the make and model of the transmitter.
- Place the circuit board on a dry, flat surface and spray the top of the board and the edges with a polyurethane coating (ITI recommends Humiseal, Type 1A27). Spray only light coats and be careful not to spray a coat so heavy that the polyurethane runs onto the masked areas.
- Between coats, allow the board to sit for approximately 5 min. After spraying, allow the board to sit until the coating no longer is sticky.
- Turn over the printed circuit board and place four light coats of polyurethane on the back surface.
- After the board is totally dry, remove the tape, replace the battery, install the board in the transmitter case, and install the transmitter.

Source: Interactive Technologies Inc. (ITI), 2266 North Second St., North St. Paul, MN 55109.

BATTERY-BACKUP PBX SYSTEMS

Wally Larman experienced a problem while attaching a digital communicator to an outgoing phone line. When he discovered that the incoming phone line was a ground-start line instead of a loop-start line, he attached his digital communicator to an unused station circuit of the PBX system. The reasoning behind this was that most modern PBXs will provide an outside line to only one station at a time.

Mutual Central Station Alarm Services Inc., New York, NY suggests that this may be an unsafe procedure because the PBX system may not have battery backup power that meets or exceeds that of the alarm equipment. The dealer also warns that, when alarm dealers connect their alarm systems to a PBX, they place the reliability of their system's communications in the hands of a PBX service company. His other concerns are the programming skills of a PBX company's technicians and whether a technician might later disconnect the alarm system from their phone system because he or she is not sure what it is.

These are legitimate concerns that must be considered before connecting an alarm communicator to a PBX station circuit. One solution is to verify that an uninterrupted power supply is used with the PBX system and to attach a label to the phone circuit at the punchdown blocks stating the purpose of that station circuit. The other solution is not to connect to the PBX in the first place. In this case, Wally Larman should go back to the owners and have them order a loop-start telephone circuit.

ALARM-CONTROLLED GUARD DOGS

Can old dogs learn new tricks? An alarm technician for American Guard Dog, Wichita, KS, believes that they can. In addition to installing and servicing alarm systems, the company also cares for guard dogs belonging to more than 100 commercial clients throughout the Wichita area. Guard dogs are dangerous and messy. Not all owners want a guard dog roaming around their furniture store, ice cream parlor, pizza shop, or warehouse full of china dolls. The dealer solves this problem by installing a solenoid on each dog pen. When an alarm is triggered, the bell output unlocks the dog pen and the dogs are released. Why not go one step further and install PIRs, wiring

Burglar Alarm Installation

them between the bell output and the solenoid (see Fig. 3–14)? This will prevent unwanted dog releases.

A QUESTION OF SOUND

Should bells and sirens sound off outside homes and businesses protected by central-station-monitored alarm systems? An alarm installer from Certified Alarm Co. of Sheffield, Alabama Inc. wants to know whether outside audibles should always be installed to UL requirements, regardless of whether the job is going to be UL certified. My personal preference is to use outside audibles at all times,

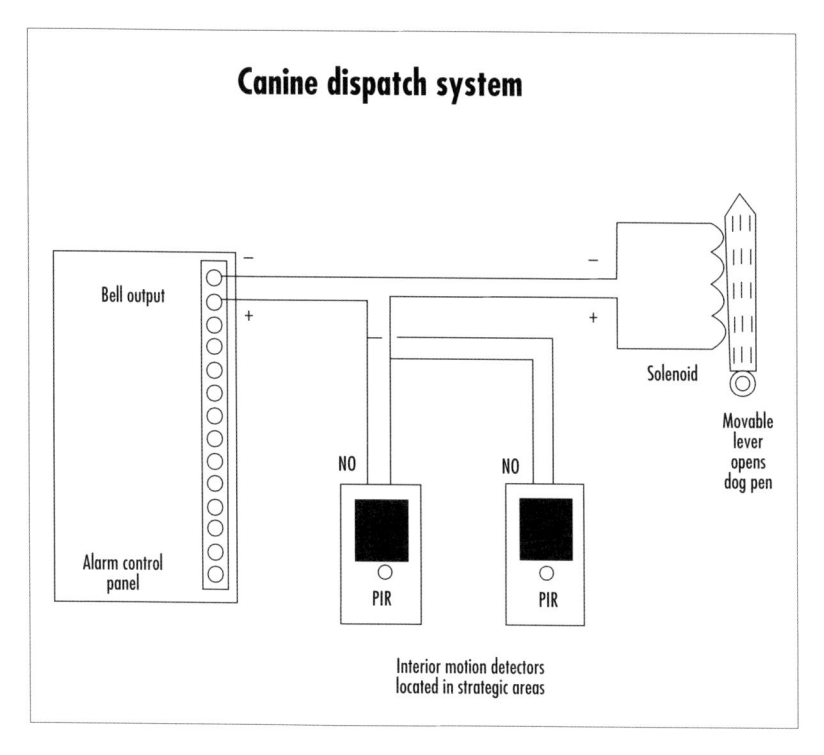

FIGURE 3–14. Both an alarm system and a PIR must trip before guard dogs are released.

especially in systems that include fire alarm components. However, some communities' ordinances ban loud noises above a certain decibel level. Also, some customers may not want to disturb their neighbors by false alarms, which shouldn't but sometimes do occur. This can be avoided by backing up central-station-connected installations with an outside siren or bell, connected in series with the NO contacts of a phone line monitor and an alarm panel's bell output (see Fig. 3–15). Some monitors continuously check phone line voltage and on-hook current. When a phone line is cut, broken, or otherwise compromised, the outside noisemaker is enabled, allowing it to operate when a burglar triggers an alarm.

AILING POWER CAUSES FALSE ALARMS

False alarms and the unstable operation and detection range of non-voltage-regulated motion detectors are symptoms of a power supply

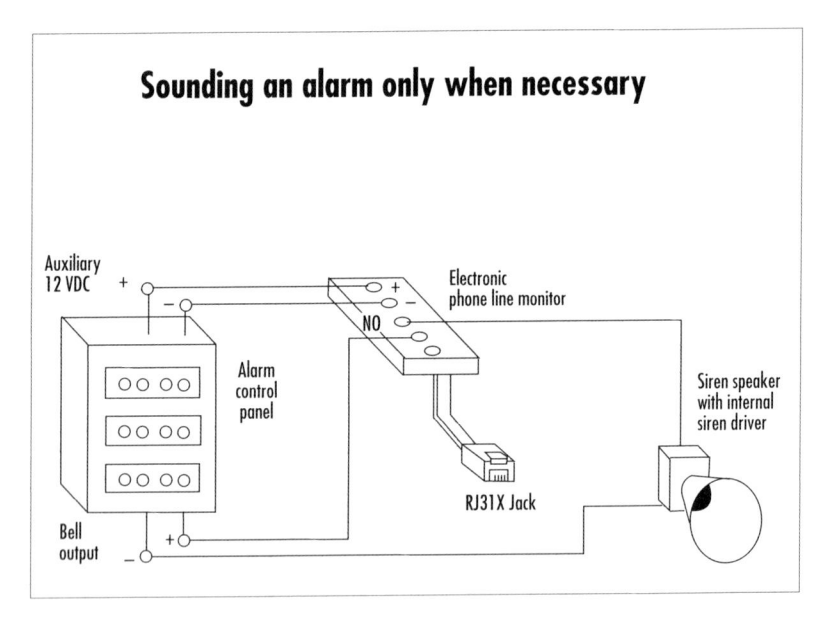

FIGURE 3–15. A phone line monitor, wired in series with an alarm panel's bell output, "enables" an outside audible when a phone line is cut or is out of order.

Burglar Alarm Installation

in trouble. Sometimes this results from a battery losing its storage capacity. For example, a number of older-model alarm panels rely on a fully charged battery to filter (smooth out) pulsating dc voltages from a rectifier. A pulsating dc voltage can cause non-regulated-motion detectors to sound a false alarm. The obvious cure for this is to replace the batteries on a regular basis. Overloading an alarm panel's auxiliary power supply also can cause problems. For example, motion detectors can draw more current in occupied areas during unarmed periods. This can starve a marginal battery, not allowing it to charge to its capacity. Then, when a power outage or brownout occurs, the auxiliary voltage rapidly drops below an acceptable level. The result is false alarms or complete alarm system failure. Adding an auxiliary power supply takes the load off of the control panel, providing more power to a starving battery.

PROVIDE ARM-ONLY CAPABILITY

Providing a janitorial service in a commercial facility, a house-watching neighbor, or a babysitter with both arming and disarming capability can be risky business. Security dealers who install today's alarm control panels, however, can provide users with an arm-only feature that provides only limited use of a burglar alarm system. The arm-only feature in most alarm control panels is inherently invoked by selecting it in programming. However, some alarm control panels, especially older models, do not offer this feature. Nick Markowitz, Markowitz Electric Protection, Verona, PA, provides his clients with this feature by connecting the normally open (NO) panic contacts in an alarm system keypad to two relays connected in tandem. Then, when someone presses the keypad panic buttons (usually * and #), power is provided to the coil in RY2 through the normally closed (NC) contact in RY1. When RY2 is energized, the NO contacts in RY2 close, prompting the alarm control panel to arm (see Fig. 3–16). Once armed, Markowitz prevents the same act from disarming the system by disconnecting the keypad panic button(s) from the coil of RY2. He does this by activating the coil in RY1 with either a programmable voltage output or a positive-armed-LED output. With RY1 energized, the NC contacts remain opened—until the system is disarmed with the proper keypad code. Installers also can use a separate arm-only keypad in place of the keypad panic switch.

Only arm for unauthorized users

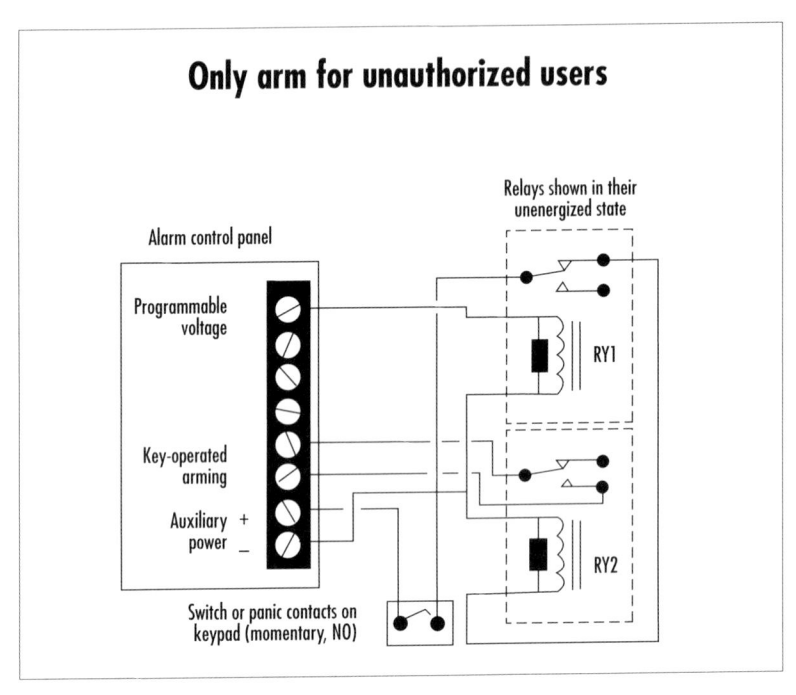

FIGURE 3–16. Nick Markowitz designed this circuit to provide arm-only capability to his commercial security systems.

PROPERLY ZONE AND RECORD YOUR ALARMS

It's vital that you properly zone a multiple-zone security system, especially in facilities larger than 40,000 sq. ft., says ACME Central Station Alarm Co., Lynbrook, NY. It is equally vital that you accurately record the zones you create. Here's ACME Central's ten-point plan for proper zoning and zone record keeping:

1. Develop a zone schedule form that includes columns for recording the system zone numbers, device type, zone response, and location of the end-of-line resistors.
2. Installers should not create zone schedules on larger jobs. This should be done by the installation supervisor.

Burglar Alarm Installation

3. Installers can create their schedules on small jobs. However, these should be reviewed by the installation supervisor after the job is completed.
4. Sales personnel must understand their objectives when zoning a system. They will create better proposals that way.
5. Draw a basic floor plan of the facility to be protected. Use symbols for the various burglar and fire devices on the drawings.
6. Before work begins on large, multiple-zone installations, go over the proposed zone schedule and floor plan with the salesperson and the installer.
7. Tape a copy of the zone schedule and floor plan inside the alarm control panel.
8. Keep a copy of the zone schedule and floor plan in the job or contract file in your office, shop, and/or central station.
9. When there are a lot of exterior or interior devices, give the subscriber a copy of the zone schedule and floor plan.
10. Instruct the subscriber to keep the schedule and floor plan in a safe place.

4
Servicing Burglar Alarms

PREVENT AC-INDUCED FALSE ALARMS

An alarm dealer asked, "I've been experiencing false alarms on a particular loop in a burglar alarm system. I've replace the detectors on the loop and I'm still having false alarms. Where do I go from here?"

The first thing to do in a case such as this is to swap the offending loop with a known good one in the same control panel. Even though this is a temporary test, you should notify the central station of the change so, in the event a false alarm is sounded, the firm will respond in the proper manner.

Then, if the false alarms follow the offending loop, check for an ac voltage on the incoming loop. To do this, remove both loop wires from the terminals on the motherboard. Using a sensitive volt/ohm meter (set it for the lowest ac setting), test each wire individually with one of the meter leads connected to earth ground at the control panel.

To reduce the level of induced ac on the loop, connect a 50-microfarad, 25-Vdc electrolytic-type capacitor (observe polarity) to each side of the loop and connect the other end of each cap to earth ground inside the control panel, says a representative for Alarmax Distributors Inc., Plymouth Meeting, PA.

SUPPRESS EMI IN CABLES

Electromagnetic interference (EMI) can cause false alarms in alarm systems and extreme noise problems in audio-related systems. The spark plugs in automobile engines, FM radio signals that emanate from commercial broadcasting antennas, hand drills, and electric motors are all EMI generators.

The best way to solve an EMI problem is to stop it at its source. However, this is not always practical. The next-best thing is to choke off the interference before it enters equipment. This can be done using snap-on EMI suppressors (see Fig. 4–1) manufactured by Fair-Rite Products Corp., Wallkill, NY.

Toroids add a large high-frequency impedance in series with wires. They block EMI from getting into electronic equipment. Toroids made of ferrite #43 introduce an impedance of 20 Ω at a frequency of 20

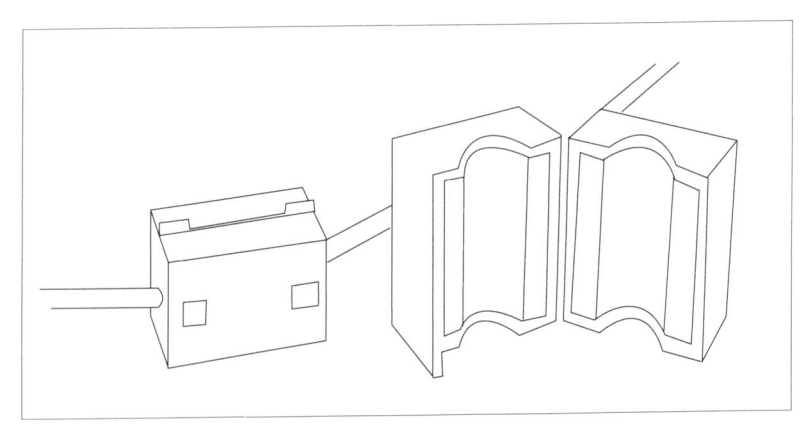

FIGURE 4–1. Snap-on toroids commonly attenuate high-frequency EMI in alarm system and audio-related cables.

Servicing Burglar Alarms 51

MHz, 40 Ω at 120 MHz, and 90 Ω at 30 GHz. For lower frequencies, toroids made of ferrite #73 attenuated radio frequency signals beginning at 10 MHz. Other frequencies are allowed to pass through a toroid normally.

You can insert more than one cable into a toroid. After the cables are in place, the toroid clamps shut over the cables (see Fig. 4–1).

For more information, contact Fair-Rite Products Corp., P.O. Box J, One Commercial Row, Wallkill, NY 12589, call (914) 895-2055, or fax (914) 895-2629.

ISOLATE PROBLEM LOOPS FROM RFI

It is not uncommon for high-power radio-frequency interference (RFI) to be the cause of unexplained false alarms. RFI can be particularly frustrating because it often appears unannounced and finding the source is very difficult.

RF signals sometimes can enter alarm panels through the external wires that connect them to security sensors, such as PIRs and door switches. When the length and capacitance of these wires form a tuned circuit near the resonant frequency of the RF signal, a voltage builds up across the tuned circuit. This results in false alarms.

When intermittent false alarms occur, establish whether RF is the problem. If conventional troubleshooting methods reveal nothing, install an SPST relay in series with the alarm circuit (see Fig. 4–2)

NC detection devices, such as switches and PIRs, should be wired in series with the relay's coil. If an EOL (end-of-line) resistor is used, it also should be removed. One side of the coil connects to the positive side of the panel's auxiliary power supply. The loose end of the alarm circuit connects to the negative side. The diode, which is inserted across the coil, suppresses high-voltage spikes that occur when switch contacts open and close.

The relay's NC contacts then are wired in series with the alarm panel's zone input, which isolates the panel's circuitry from RF sources. If necessary, place the EOL resistor in series with the relay contacts and alarm circuit.

Isolate alarm loops from RF

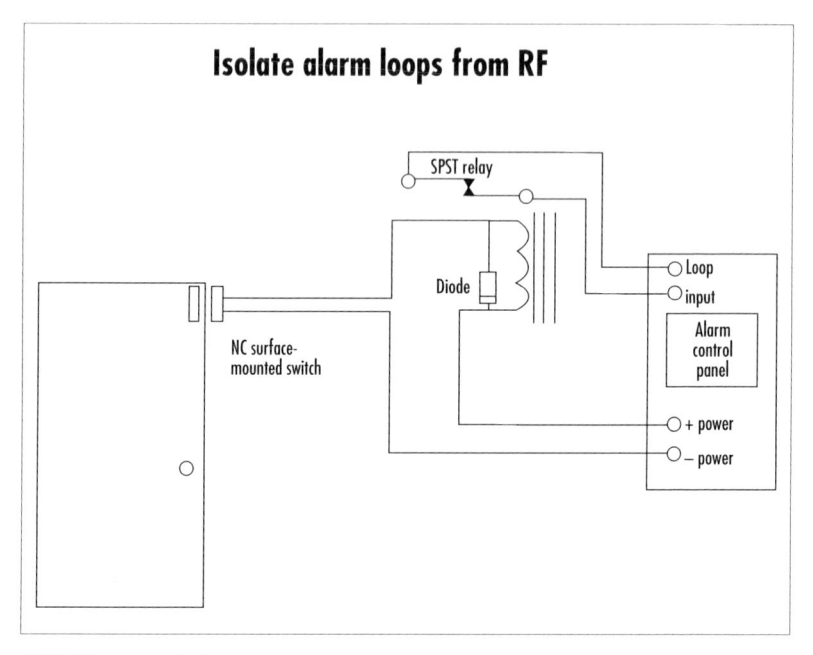

FIGURE 4–2. Radio frequency interferences can be isolated from alarm panel zone-input circuits by using conventional relays. Match the voltage of a relay coil to an alarm panel's auxiliary power output. Then use Ohm's law to verify that the specification of the switch will safely manage the current that will flow through the powered relay circuit.

SHORTING OUT YOUR RFI PROBLEMS

Alarm panels are susceptible to false alarms when lengthy zone wires act as antennas. RFI often is directed back to an alarm panel. Here it is rectified and amplified by nonlinear electronic components.

The first step in solving an RFI problem is to determine whether RFI is the culprit. To do this, either remove the sensor circuit, placing another EOL across the panel's zone input, or isolate the zone input from the loop using the method discussed in Kinks & Hints, "Isolate Problem Loops from RF" (August 1991, p. 57). If the false alarms stop, then RFI is probably at fault.

Servicing Burglar Alarms   53

You can minimize the effect of RFI by installing an easy-to-build RFI filter circuit (see Fig. 4–3). The most simple filter (top zone input) is a single capacitor (C) placed in parallel across a zone's input. The capacitor creates a low-impedance path for RFI, shorting the majority of the interference past the input of the zone.

Single-section tuned LC circuits (middle zone input) create a high-impedance path to RFI but only a minimal amount of resistance to an

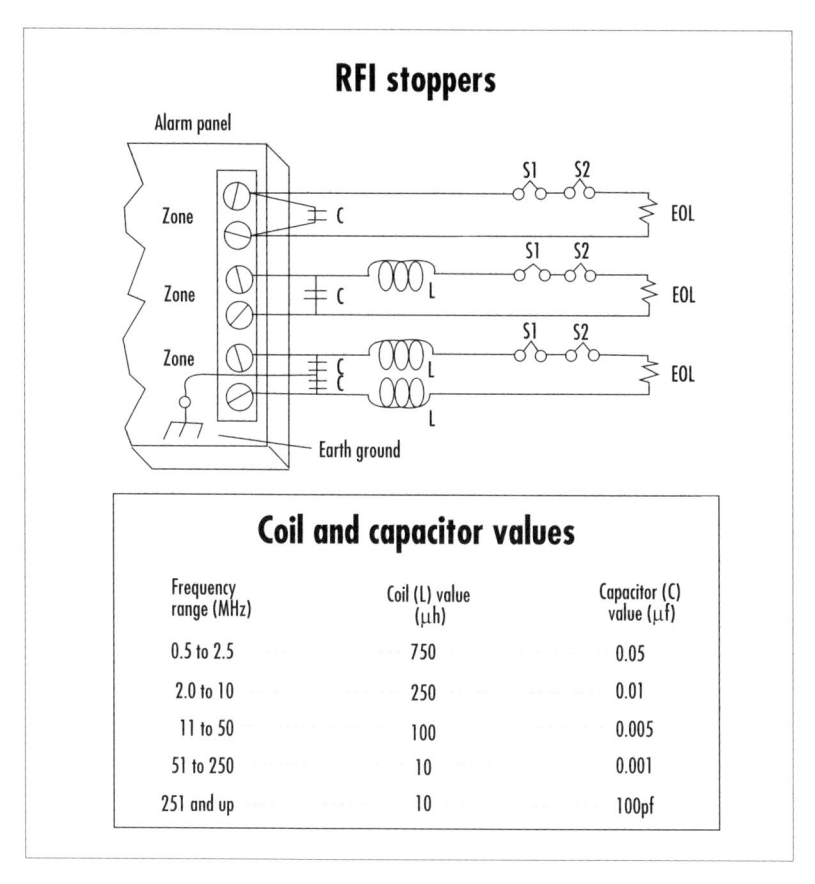

RFI stoppers

Coil and capacitor values

Frequency range (MHz)	Coil (L) value (μh)	Capacitor (C) value (μf)
0.5 to 2.5	750	0.05
2.0 to 10	250	0.01
11 to 50	100	0.005
51 to 250	10	0.001
251 and up	10	100pf

FIGURE 4–3.   RFI filter circuit.

alarm panel's dc current. Double-section filters (bottom zone input) provide even better RFI filtering, but they require good earth grounds.

The single- and double-section filters must be constructed; however, before beginning, try to determine the RFI's source. One way to do this is to use an expensive oscilloscope. The most affordable way, however, is to observe the affected alarm panel's environment.

For example, a large hospital in the vicinity may have diathermy machines that generate RFI (27 MHz). Nearby restaurants, another example, could have microwave ovens that produce RFI (2 GHz). Also, when an affected system is near a marina, marine radios can generate RFI (156.25 MHz to 162.0125 MHz).

The values of L and C (see the chart in Fig. 4–3) depend on the frequency of the RFI. If the frequency of the RFI cannot be established, you will have to take an educated guess, build a filter, and try it. You may have to build several filters until the correct frequency range is found.

To construct an RFI filter, solder the components on a predrilled pc-board or connect the lead-in wires of each component directly to the panel and external zone.

RADIO FREQUENCY INTERFERENCE SOURCES

In the section "Shorting out Your RFI Problems" you learned how to minimize RFI by building capacitor/inductor filters. Before you can start building filters, however, you need to know the frequency of the RFI that's causing your problems.

One way to do this is to visually search for nearby RFI sources. For example, FM (frequency modulation), AM (amplitude modulation), and TV broadcasting stations can cause RFI problems. Microwave towers and hospital diathermy machines also can cause RFI problems.

If a visual inspection of the surrounding area fails to uncover your RFI source, you can turn to a handheld programmable scanner. Although expensive, these devices can recoup your investment in time saved by quickly tuning in on RFI sources. Many of today's scanners can identify frequencies between 30 MHz and 1 GHz. This encompasses almost every frequency likely to cause alarm system interference problems.

After the RFI source has been identified, use Fig. 4-4 to determine the frequency causing your problems. Then, using the capacitor/inductance chart in Fig. 4-3, you can build an RFI filter.

TROUBLESHOOT DEFECTIVE SWITCHES

Normally closed door and window switches sometimes fail in an open-circuit mode. When one out of several goes bad, finding it can be difficult.

First, the current flowing through D5 (see Fig. 4-5) is the same as the current that flows through the EOL, D1, D2, and so on. Therefore, with D5 open, for example, the voltage between the positive and negative (Zone 1) terminals will appear across D5. This voltage can be measured by placing the black meter probe on the negative side of the switch (left side of D5) and the red meter probe on the positive (right) side. When a voltage reading appears on the meter's LCD readout, D5 is bad or the door is open.

To find an open NC switch when there are more than one series-connected switches, place your meter's test probes across the contacts of each switch, one at a time. When the needle, or LCD readout, registers a voltage reading, you have found the bad switch.

HOW TO TROUBLESHOOT SWINGERS

Troubleshooting intermittent false alarms can be time consuming and difficult, says Jim Williams, installation foreman for Home Services Inc., Canton, OH. He says that he uses a circuit tester and a set of walkie-talkies to determine the integrity of switches and other normally closed devices.

Williams uses electrical tape to hold down the transmit button on one of the walkie-talkies. This is done by holding down the transmit button and circling the walkie-talkie with the electrical tape. Next, he disconnects the problem circuit from the alarm control panel and clips the loop tester to it. Then, he places the loop tester on top of the transmitting walkie-talkie at the alarm control panel with the tester's piezo sounder face down on the walkie-talkie's speaker.

Williams tests magnetic switches by gently tapping each switch housing with the butt end of a screwdriver. If the contact inside the

Sources of RFI

AM broadcast:	0.05 to 1.6 MHz
Amateur radio:	1.8 to 2 3.5 to 4 7 to 7.3 14 to 14.35 21 to 21.45 28 to 29.7 50 to 54 144 to 148 220 to 225 420 to 450 1,215 to 1,300 2,300 to 2,400
Business radio:	152.27 to 159.21
Citizen's band:	26.96 to 27.41
FM broadcast:	88 to 108
Garage doors:	27 130 to 500
Induction heating:	2 to 30
Marine radio:	156.25 to 162.0125
Medical diathermy:	27
Microwave ovens:	2,000
Military radio:	2 to 150.5
Police/fire department:	1.61 to 1.73
Public utility:	2.366 to 2.49 2.804 to 2.812 5.135 to 5.195 7.48 to 7.935 33.12 to 33.90 37.02 to 46.5 72 to 76 152.27 to 159.21 450 to 470
Radar:	1,200 to 7,000
Shortwave:	5.96 to 6.2 9.5 to 9.775 11.7 to 11.975 15.1 to 15.45 21.45 to 21.75 25.6 to 26.1
Television:	54 to 88 174 to 225 470 to 890

FIGURE 4–4. Use this table of possible sources of electromagnetic interferences and their corresponding frequencies to determine the inductance/capacitance values necessary when building simple filters. All frequencies are in MegaHertz (MHz).

Servicing Burglar Alarms 57

switch housing is bad or near bad, it will trigger the tester and he'll hear it in the receiving walkie-talkie.

Not only is Williams able to test the holding ability of a switch under stress, he also can listen in to the quality of the contacts inside them. For example, good contacts rarely produce static in the receiving walkie-talkie when Williams opens and closes them. Contacts that are bad or going bad, however, usually generate static in the speaker of the receiving walkie-talkie.

To determine the quality of normally closed sensors, for example, William goes from sensor to sensor, striking each one with a screwdriver. He also induces vibration in the adjacent structure to see how each device reacts. This technique usually reveals the true condition of each device according to the amount of static it generates in the receiving walkie-talkie.

This same arrangement also allows him to trace wires in new alarm systems and to check the closed/open position of doors, windows, and other devices in a protected facility.

PREVENT FOIL-TAPE FALSE ALARMS

Dealers that experience intermittent false alarms in commercial facilities know how frustrating it is to fix the foil on windows. One

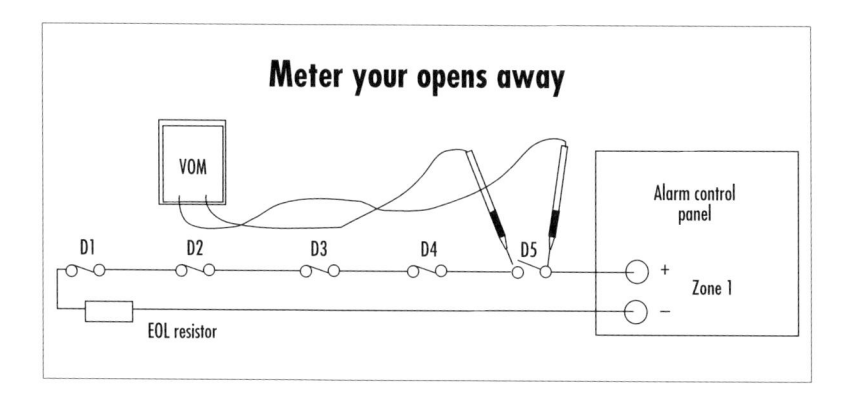

Meter your opens away

[Diagram: VOM connected across switches D1–D5 in series with EOL resistor, leading to Alarm control panel Zone 1 terminals]

FIGURE 4–5. Use a digital or analog meter to check the integrity of series-connected switches.

way to do this is to attach the negative (black) probe of a voltmeter to the negative side of a foil circuit and measure the voltage over the entire strip of foil. You can do this by placing the red (positive) probe at position 1 (see Fig. 4–6) and writing the voltage on a piece of paper for future reference.

Next, measure the voltage along the entire length of tape at equal intervals (positions 2, 3, 4, etc.). The sum of all the voltages should equal the reference voltage at position 1. Thus, the voltage readings taken at each position should decrease incrementally (providing the distance between each position is about the same) by the same amount.

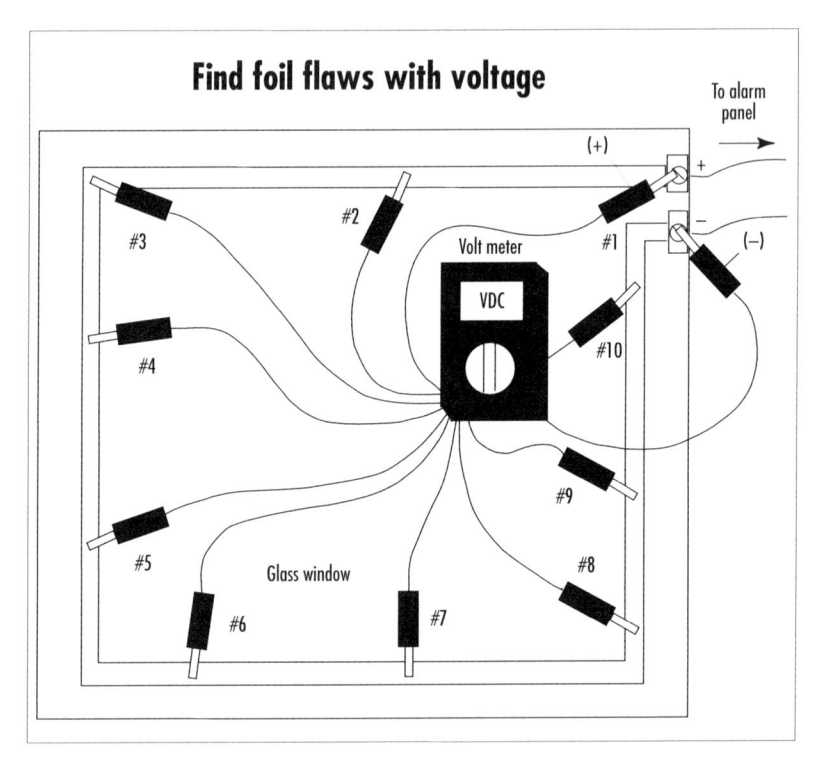

FIGURE 4–6. Security dealers can locate foil-tape problems by clipping the negative probe of a voltmeter to the negative side of the foil tape (block) and measuring the voltage at equal distances along the foil tape.

Servicing Burglar Alarms

A disproportionate decrease in voltage anywhere along the foil may indicate a badly oxidized connection. The bad connection usually can be found between the point where the disproportionate voltage was taken and the meter reading before it.

STOP BAFFLING MICROWAVE FALSE ALARMS

When microwave motion sensors are installed too close above a suspended ceiling (see Fig. 4–7), false alarms can occur. This is due to signal bounce-back. Although signal bounce-back false alarms are somewhat rare, when they occur, alarm technicians can easily overlook their real cause, resulting in multiple false alarms from the same initiating zone. This is because this kind of false alarm is random and its cause is not apparent.

All microwave motion sensors are subject to this kind of false alarm, but recent models are not as prone. This is because improvements in signal processing enable technicians to filter out most of the effect of signal bounce-back and structural vibration. Older, vintage models, however, are more susceptible.

One way to solve this problem when it occurs is to install or move a suspect microwave motion sensor so it is 12 to 24 in. above the suspended ceiling tile. In addition, technicians should carefully adjust microwave sensors to help minimize the effect of signal bounce-back.

SOLVE THOSE DIFFICULT FALSE ALARM PROBLEMS

Your most difficult false alarm problems can be solved using a combination of four PIRs. "It works on alarm panels equipped with a common negative [between zone inputs and a power supply]," says a representative of Independent Alarms & Monitoring, Middleburgh, NY.

This circuit offers false-alarm protection by using a series/parallel circuit (see Fig. 4–8). Two adjacent closed-circuit PIRs must trip before an alarm will sound. PIRs 1 and 4, 1 and 2, 3 and 4, and 3 and 2 will sound an alarm. Detectors 1 and 3 and 2 and 4 will not.

Another benefit of this circuit is it requires only inexpensive four-conductor cable to operate. If using this technique on an alarm

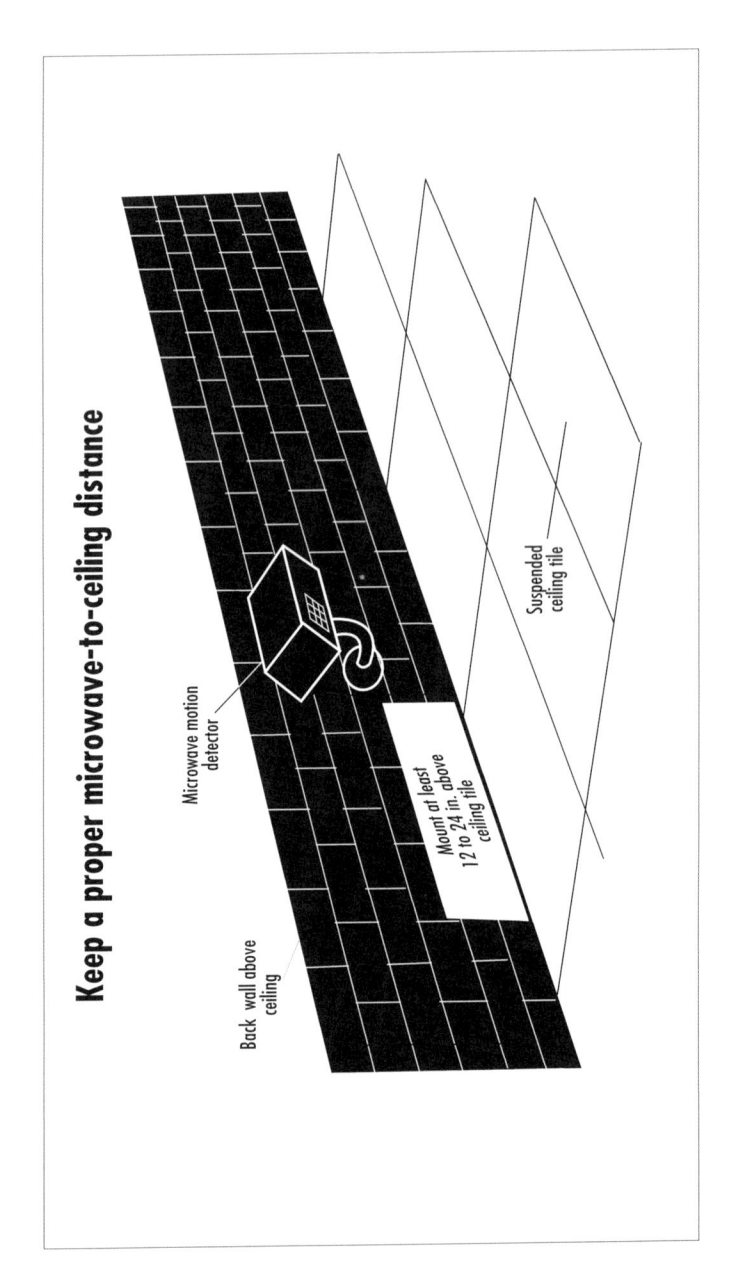

FIGURE 4–7. Install microwave motion detectors between 12 and 24 in. above a suspended ceiling to prevent false alarms from signal bounce-back.

False-alarm-free PIRS

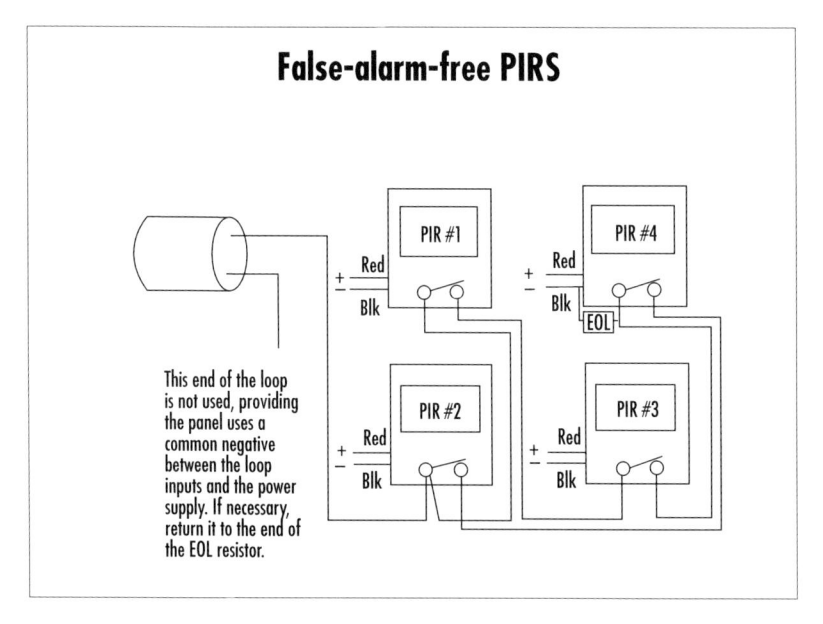

[Diagram showing power supply connected to four PIRs (#1, #2, #3, #4) wired with Red (+) and Blk (-) connections, with an EOL resistor. Note on diagram: "This end of the loop is not used, providing the panel uses a common negative between the loop inputs and the power supply. If necessary, return it to the end of the EOL resistor."]

FIGURE 4–8. Solve those difficult false alarm problems by using two adjacent PIRs. Both detectors must detect simultaneously for an alarm to occur.

panel with isolated zone inputs, use five-conductor cable to return the EOL to the negative terminal of the zone input.

FIX TROUBLESOME TRANSMITTERS

A security dealer asked, "Why do some of my wireless alarm systems experience problems with supervisory and alarm status signals reaching the receiver? Is there something I can do that will increase the likelihood that these signals will be received?"

There are two ways to solve the on-today, off-tomorrow operation that security dealers experience with some short-range radio transmitters—reposition them or relocate them.

The first thing to do is reposition the troublesome transmitters so the polar axis of the transmitter's antenna is nearly the same as the receiver's antenna (see Fig. 4–9).

Proper polarization is important because when the polar patterns of both antennas are comparable, more electromagnetic lines of force from the transmitter will intercept the receiver's antenna. The result almost always will be better reception and reliable service over the life of the alarm system.

When it comes to the polar alignment of antennas, physical repositioning is not always enough. This is especially true when the signals

Transmitter Fixes

Reposition Transmitters to Realign the Antennas

Original   90°   180°

Relocate Transmitters to Avoid Wave Nulls

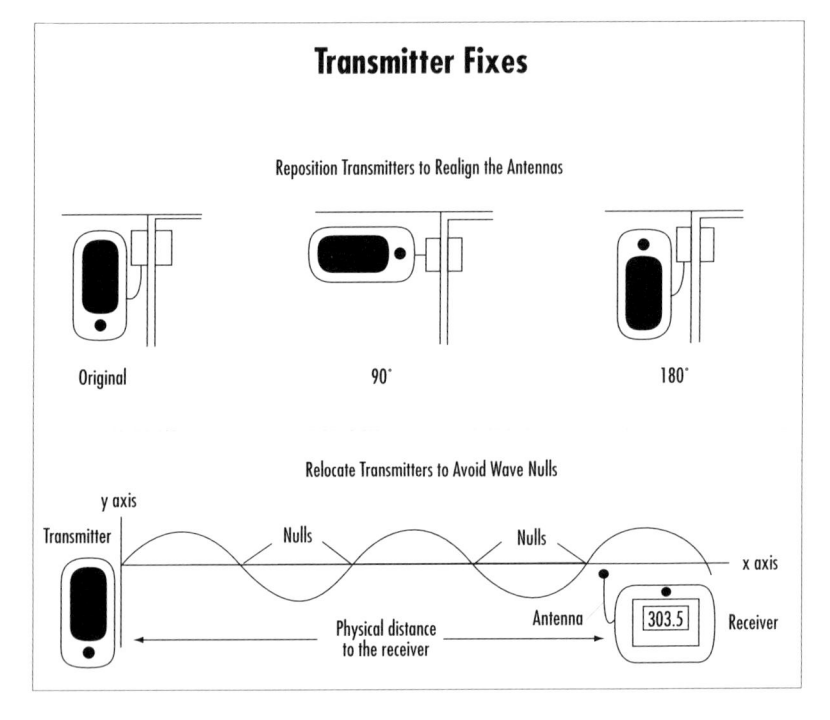

FIGURE 4–9. Security dealers should reposition troublesome transmitters to increase the amount of radio energy available at the receiver's antenna. After repositioning, the dealers should then conduct a performance test at 90° and at 180° (top). Relocating troublesome transmitters sometimes is necessary so the receiving antenna is not physically located in one of the signal's nulls. A null occurs where the signal crosses the zero-potential, x axis (bottom).

Servicing Burglar Alarms

from a transmitter are scattered in the environment, causing the polar pattern of the received signal to shift.

In this case, RF scatter occurs when radio signals bounce off of objects in the environment, causing the reflected radio wave and not the direct signal to reach the receiving antenna. Because the polarization pattern of the reflected carrier wave is changed, the physical axis of the transmitter antenna may appear to be different than the receiver's.

The second way to deal with troublesome transmitters is to relocate them. The problem associated with relocation actually is from nulls in the transmitter's signal. In physical terms, nulls are spatial locations where the energy of a transmitted radio wave is at a minimum. With the wavelength of the signal in mind, this is the spot where the radio-frequency energy crosses the zero-potential, x axis (see Fig. 4–9).

Some wireless systems come with dual antennas, which increase the likelihood that at least one of the antennas is physically at a point on either side of the null where the amplitude of the wave is rising above or falling below the x axis. For those systems with only one antenna, the dealer may find it necessary to physically move the transmitter closer to or farther from the receiver. In either case, only a fraction of an inch usually is all it takes.

BAD COMMUNICATOR CONNECTION

Dealers have reported experiencing telco-related problems after local water department workers install an automatic meter-reading device on the telephone line.

This problem is reminiscent of that when alarm dealers and telco installers differed in opinion on how these infamous jacks should be wired. After several repeat visits to the job site and heated discussions with telco service representatives over the telephone, alarm dealers finally took the time to show them how to install their own equipment.

If you are experiencing this kind of problem, maybe it's worth taking the time now to show the water department workers how it's done. The Pennsylvania Fire & Burglar Alarm Association may take the initiative in its locality and invite water department employees to learn about RJ31X jacks.

5

Central Station Monitor

DIAL-OUT DELAY AVOIDS FALSE DISPATCHES

False dispatches can be avoided if you program a 10-sec communicator dial-out delay in modern alarm control panels equipped with an integral digital communicator or install an electronic timer in older panels equipped with a slaved digital communicator. This also delays the activation of inside and outside bells, sirens, and strobe lights.

PROGRAMMING ERROR CAUSES FALSE DISPATCHES

False alarms are one of this industry's most frustrating problems, but dispatching on the wrong account is even worse. One technician at API Alarm Inc., Toronto, Canada, solved a false dispatch problem in his regional central station. When some of the alarm systems monitored tried to send restoral or test-timer signals, the operators received the wrong account number and condition code. Consequently, they dispatched authorities on the wrong subscriber accounts.

The dealer's investigation revealed that these mistakes were due to simple programming oversights. All of the dealers involved were using control/communicator panels that require two separate program entries for restoral and test-timer signals.

For example, these programs typically ask if the dealer wants to enable restorals and test-timer functions. The condition codes then are entered at a prompt later in the program. Dealers simply were forgetting to enter their condition codes. The result of the oversight was the transmission of false account numbers and condition codes.

For example, a burglar alarm drop (condition code 3) at subscriber A's premises (account number 1234) should result in the digital transmission of 1234-3. If the dealer entered a "yes" at the "enable" prompt but failed to enter a condition code later, the central station receiver would interpret the ensuing signals as 123-4. If the code 4 received requires a fire response, subscriber B would soon receive a visit from the local fire department.

A good understanding of central station receivers is necessary to understand how this fluke can happen. Under normal conditions, when a four-digit account number followed by a one-digit condition code (4/1 format) is transmitted, a digital receiver will interpret the five-digit transmission in 4/1. Or, when a three-digit account number and a one-digit condition code is received (3/1 format), most receivers will interpret it in 3/1.

However, when a dealer forgets to program a condition code, the communicator first transmits the account number (1234) and then looks for the condition data. In this case, the communicator will transmit exactly what it sees: nothing. Since receiver intelligence is limited, the transmission is interpreted as account number 123 with a condition code of 4.

So far this particular dealer has experienced this problem only in 10- and 20-baud formats (that's all the central station accepts). The dealer is not sure if the problem still occurs using BFSK, super-fast, or some other format.

He also notes that this problem may not occur in every control communicator that uses two program locations for restorals and test timers. Some manufacturers already have included filters in their software to force the program to look for data in a condition location. If it is not there, the communicator aborts before it dials out.

Central Station Monitor

If you use a control/communicator that uses this type of software, you can prevent the problem by programming condition codes for restorals and test timers. Manufacturers also should include filters in their panel/communicator software.

BETTER PROTECTION FOR RJ31X JACKS

Burglars can stop an alarm system from reporting to a central station by simply lifting the handset of a telephone inside a protected premises. They also can dial bogus numbers while the communicator tries to dial out to a central station over the conventional telephone network.

The answer to this potential problem is to install an RJ31X (series) jack that, when installed properly, routes an incoming telephone line to a control panel's digital communicator before connecting it to the inside telephones. This enables the digital communicator to disconnect a defective communicator/alarm system that permanently seizes a customer's inside telephone line.

Installers can either order the installation of an RJ31X jack from the local telephone company or install one themselves. To order one, for example, call the telephone company and give it the FCC registration number and the ringer equivalence of the digital communicator then wire an RJ31X male plug to the alarm control panel. Connect the brown and gray wires to the outside phone line terminals, and the green and blue wires to the inside telephone terminals (see Fig. 5–1).

To install a RJ31X jack, first choose a mounting location that is secure from visitors yet accessible to the customer. In the past, some alarm installers placed them inside the alarm control panels, where their customers and telephone repair people could not find them.

Second, connect the incoming phone line to terminals 4 (tip) and 5 (ring), and the inside telephones to terminals 1 (tip) and 8 (ring). This enables the digital communicator to disconnect the inside telephones while it communicates with a central station.

In addition, install a jumper wire between terminals 2 and 7 and then connect the blue and orange wire from the harness to a 24-hour tamper circuit inside the alarm control panel. This prevents people

Making sense of the series jack

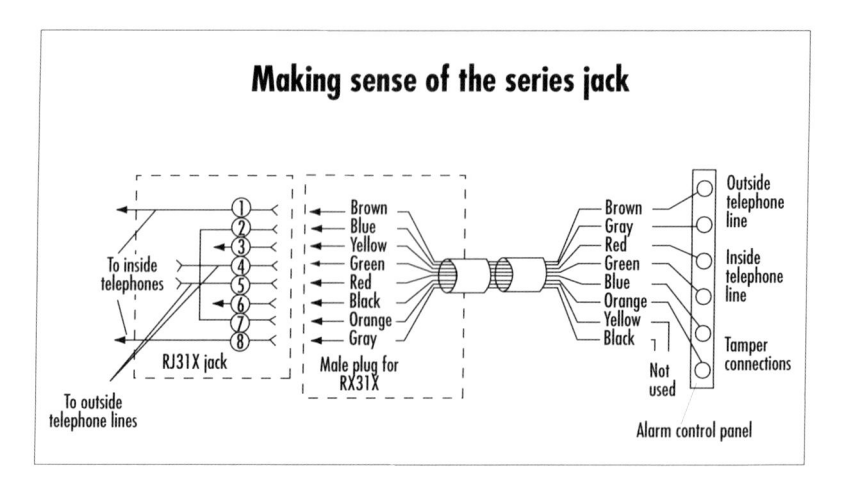

[Diagram showing RJ31X jack wiring: pins 1-8 connecting to inside telephones and outside telephone lines, with male plug for RX31X showing Brown, Blue, Yellow, Green, Red, Black, Orange, Gray wires connecting to Alarm control panel with Outside telephone line, Inside telephone line, and Tamper connections (Brown, Gray, Red, Green, Blue, Orange, Yellow, Black, Not used).]

FIGURE 5–1. RJ31X jacks route outside telephone lines to digital communicators before they go to the inside telephones.

from deliberately or inadvertently removing the male plug from the RJ31X jack. Otherwise, disconnection will render the digital communicator inoperative.

6

Do-It-Yourself Circuits

PROTECT YOUR REED SWITCHES

When surge currents are induced in detection loops, reed switches can become welded together. Meter readings and loop supervisory currents usually cannot detect this condition. This leads to undetected burglaries, financial loss, and possible legal action on the part of an unhappy customer.

A representative of Nascom Inc., Newberg, OR, says that installing 0.1 A fast-acting fuses in series with perimeter reed switches will solve the problem.

In-line fuses will burn out when currents above 0.1 A occur, says Nascom. This either prevents damage to a switch or causes an open connection between a damaged switch and an alarm panel, which, in turn, prevents end users from arming the system without first attending to the problem.

In-line fuse holders and fuses will work, and they are more convenient when fuse replacement is necessary. But the low current of today's microprocessor-based alarm panels, combined with the dissimilar metals used in fuses and fuse-holder contacts, can cause oxidation. This results in resistance buildup, which causes false alarms.

Pig-tail fuses work better because their lead wires are soldered to the end caps of the fuses. This eliminates oxidation, resistance buildup, and false alarms. Heat-shrink tubing (see Fig. 6–1) should be installed over the fuse and soldered connections, then heated until it forms a sealed covering.

ELIMINATE BELL NOISE

Bells that operate on low-voltage dc can inject noise into the cable that connects it to an alarm system. This can cause electronic components on the same circuit to malfunction.

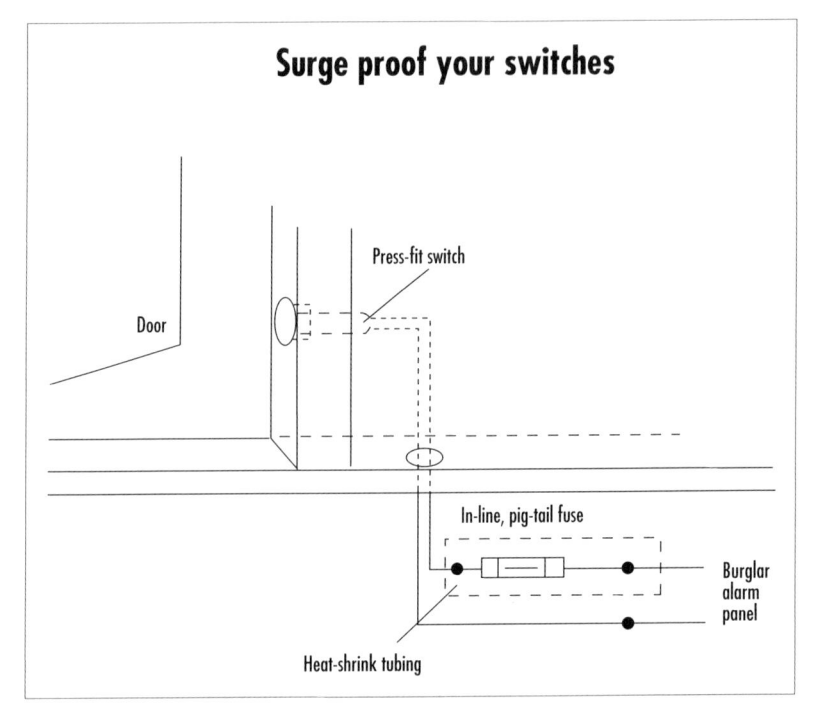

FIGURE 6–1. An in-line, pig-tail, 0.1 A fast-acting fuse is installed in series with a sensitive reed switch to protect it from surges. This opens the NC alarm circuit, preventing the arming of the system until the problem is fixed.

Do-It-Yourself Circuits

Bell noise is the result of a dc current as it abruptly stops and starts inside a bell. Each time this happens, the magnetic energy saved in a bell's coil is released to the circuit in the form of a potential difference, commonly referred to as a back electromotive force (BEMF). The problem occurs because the BEMF essentially modulates the dc current, injecting a voltage spike into the same low-voltage line that feeds other devices.

Because the polarity of the BEMF is opposite that of the dc source, dealers should connect a diode across the bell's input (see Fig. 6–2). Connect the diode's cathode (negative) to the positive side of the bell and the anode (positive) to the bell's negative side. In this configuration, the diode offers maximum opposition to the dc current while acting as a short to the BEMF.

PUT CAPACITORS TO WORK

Capacitors are important to alarm installers. They use "storage plates" to store and discharge electrical charges. When an SPDT switch connects a power source to a capacitor (see Fig. 6–3), electrons

Eliminate voltage spikes

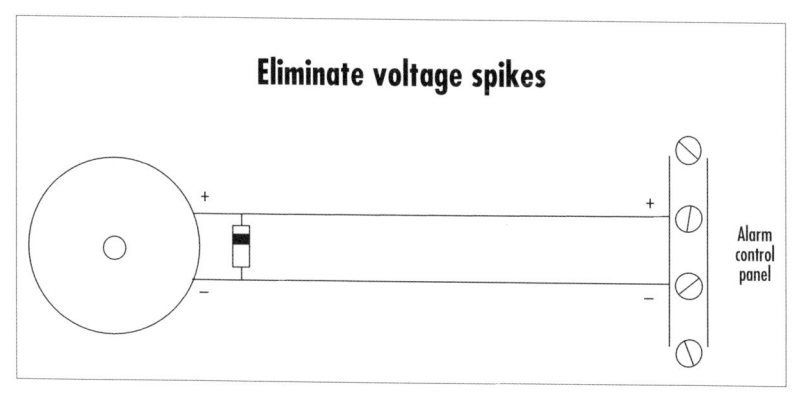

FIGURE 6–2. A collapsing magnetic field in a bell creates a back electromotive force that induces spikes in the cable that feeds it. To reduce spiking, connect a 2.5 A, 1000 PIV diode across the bell's input.

Capacitors store and discharge electricity

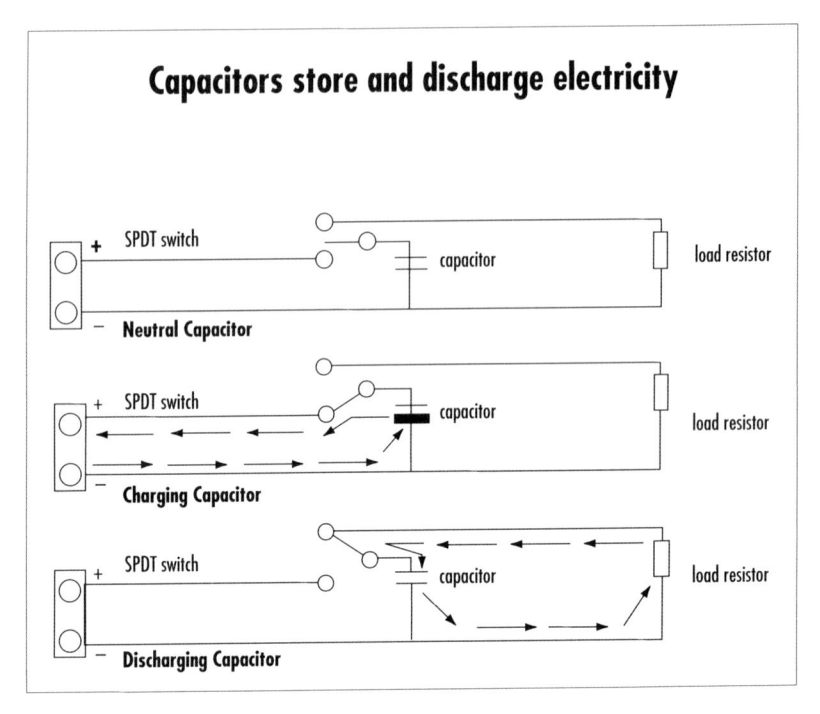

FIGURE 6–3. Capacitors, like batteries, store electricity. Unlike batteries, capacitors discharge it in one shot.

begin flowing from the negative terminal of the power source to the capacitor's bottom plate. Electrons, already stored on the top plate, are repelled by the growing negative charge on the bottom plate. This action forces the top plate electrons toward the positive terminal of the power source.

Placing a load resistor over a charged capacitor reverses this process, allowing electrons to flow from the capacitor's bottom plate, through the load resistor, to the top plate of the capacitor. When the charges on both plates become equal, the current flow slows to a stop.

You can prevent potential fires in high-output siren systems by using capacitors to block a dc voltage (see Fig. 6–4). Capacitors can be installed over NC alarm loops to filter out radio frequency signals

Do-It-Yourself Circuits

Putting capacitors to good use

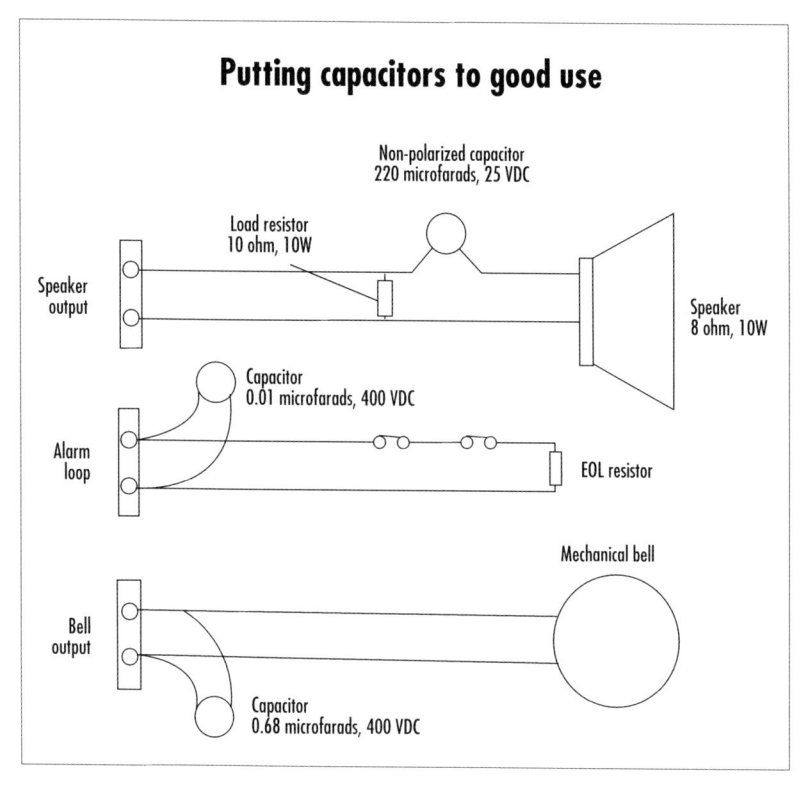

FIGURE 6–4. Use capacitors in speaker circuits to block high-current dc, in alarm loops to shunt RF interference, and in alarm output circuits to short out high-voltage spikes.

that cause false alarms as well. They also are useful in indicating circuits where electromechanical bells inject voltage spikes into alarm panels, which can interfere with the proper operation of microprocessor-based alarms.

SWITCH PLATE FOOLERY

"Why not put outdoor abandoned keyswitches to good use," Nick Markowitz, owner of Markowitz Electric Protection, Verona, PA, asks. "Here's a way to stop an alarm defeater in his tracks." Markowitz attaches unused keyswitches to 24-hour alarm circuits.

Round tubular outside-mounted keyswitches once were used to shunt door switches, large portions of alarm systems and to arm and disarm entire security systems.

Keyswitches often are found outside in doors, door frames, and switch plates. Trim-style switch plates frequently were mounted in the metal frames of doors and windows. Now, they are being replaced by an interior-mounted LED indicator or LCD textual touchpads.

When upgrading old alarm systems, it is important to either remove or disconnect outdoor keyswitches. They become security risks when they remain parallel-connected to security sensors. Burglars have to circumvent only these outside keyswitches to gain entrance to buildings.

Keyswitches often are circumvented by burglars equipped with portable drills and circular hole saws. Drilling and sawing, burglars can destroy the hardened-steel pins that secure these moveable mechanisms. Once the switches are open, they easily can turn the center plugs using a needle-nose pliers. This action changes the electrical state of a small switch installed in the back of the keyswitch mechanism.

Markowitz also lights up the red LEDs found on most abandoned switch plates by wiring them (see Fig. 6–5) to the auxiliary power output of an alarm panel. He uses an in-line fuse to protect the panel's power supply from burglars who try to short these LEDs' electrical connections.

Lit LEDs usually give switch plates a look of authenticity. However, when more knowledgeable burglars take the time to case their objective, especially when a business is open for business, a more persuasive appearance usually is necessary. For example, if a switch plate's red LED is lit all of the time, burglars may suspect that someone is attempting to fool them.

For a more convincing appearance, connect these red LEDs to a panel's "armed" voltage output. This further reinforces the burglars' conviction that circumventing the keyswitch will gain them entrance.

Not all modern alarm panels, however, are equipped with "armed" voltage (status) outputs. Usually the remedy is to connect

Do-It-Yourself Circuits

Don't waste outdoor switchplates

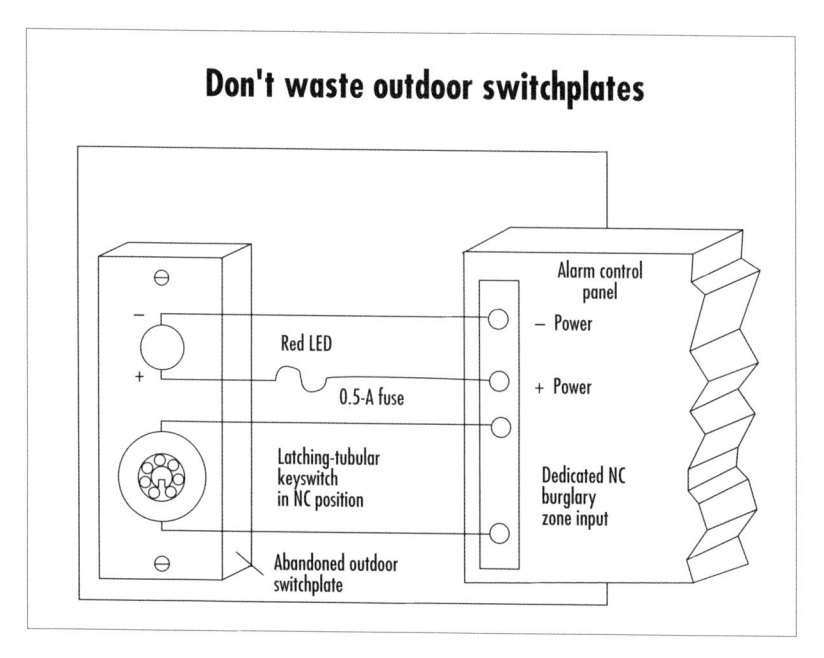

[Diagram showing an outdoor switchplate with Red LED, 0.5-A fuse, Latching-tubular keyswitch in NC position, and Abandoned outdoor switchplate, connected to an Alarm control panel with − Power, + Power, and Dedicated NC burglary zone input.]

FIGURE 6–5. Abandoned keyswitches and status LEDs can be used to detect burglars before they enter buildings.

a switch plate's LED to the panel's auxiliary power supply, using the NO contacts of a programmable auxiliary relay.

LEDs and keyswitches on most switch plates can be connected to a panel using the four-conductor cable, usually already in place. Two wires are used to power the LED. The other two connect the keyswitch to a panel's zone input. If the zone is equipped with an EOL resistor, be sure to install it at the outside keyswitch for supervision. Burglars who now attempt to circumvent an outside keyswitch will be discovered before entering a building.

ADD VOLTAGE REGULATION TO YOUR ALARMS

Smoke detectors, PIRs, ultrasonics, and glass-breakage sensors, for example, work better when they are powered by regulated voltage sources. Regulators help to eliminate ac ripples in dc outputs. Fluctuating voltage sources also can alter the detection field of motion detectors and can cause dc-operated relays to drop out, causing false alarms.

Fortunately, most control panels today have solid-state voltage regulators. However, older panels may not have one or have one that already has reached its capacity.

Voltage regulators can be added to any alarm panel with an unregulated auxiliary output. You also can build a "variable-voltage regulator" for under $5, using Radio Shack parts, and add it to an alarm panel. The variable-voltage regulator produces 1.2 to 37 Vdc with an input voltage of up to 40 Vdc. The variable resistor, R1, (see Fig. 6–6) adjusts the regulator's output voltage.

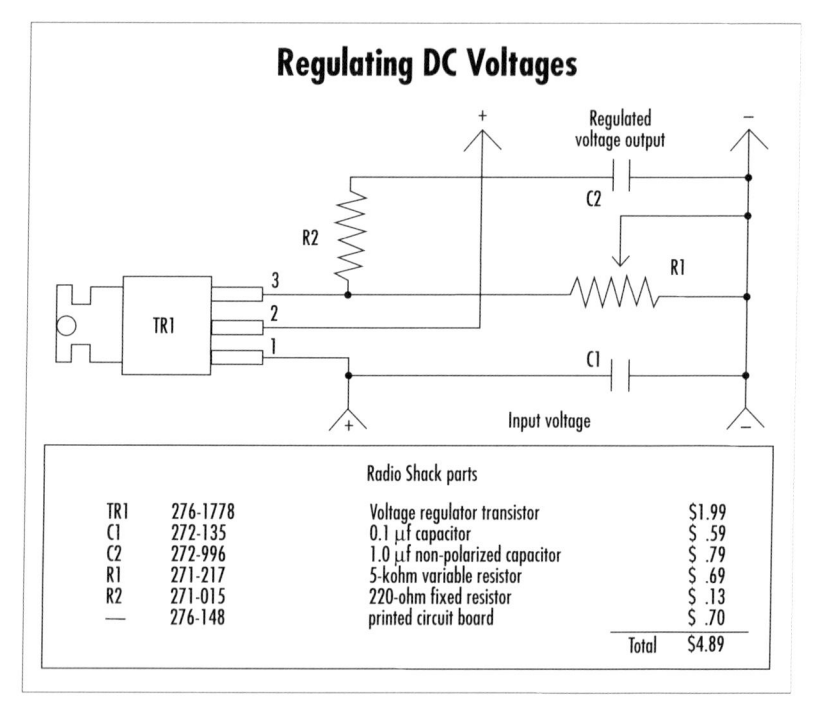

Regulating DC Voltages

Radio Shack parts			
TR1	276-1778	Voltage regulator transistor	$1.99
C1	272-135	0.1 µf capacitor	$ .59
C2	272-996	1.0 µf non-polarized capacitor	$ .79
R1	271-217	5-kohm variable resistor	$ .69
R2	271-015	220-ohm fixed resistor	$ .13
—	276-148	printed circuit board	$ .70
		Total	$4.89

FIGURE 6–6. The solid-state parts used in the variable-voltage regulator are mounted and soldered on a perforated pc-board using Radio Shack parts and a low-wattage soldering iron with 60/40 (lead/tin) rosin-core solder.

Do-It-Yourself Circuits

A low-wattage soldering iron, 60/40 rosin solder, a perforated circuit board, and a handful of parts are all that is needed.

Fixed-voltage regulators also can be built by first adjusting R1 until the desired voltage is obtained. Then measure the resistance across R1 and replace it with a 0.5 W fixed resistor of the same value. Otherwise, you can use silicon on the wheel and back plate of R1 to keep it from moving.

MATCH YOUR SPEAKERS AND DRIVERS

Siren speakers and drivers commonly are rated in ohms (Ω) and watts. For optimum efficiency, these ratings must match. For example, the 4-Ω, 10-W speaker in Fig. 6–7 will work fine when it is connected to a 4-Ω, 10-W driver. However, the driver or the speaker can fail when connected to an 8-Ω/20-W driver.

One solution is to connect a second speaker to the driver. For example, the parallel-connected speakers (see Fig. 6–7) have an equivalent resistance of 2 Ω, which is too low for this driver. This value was computed by dividing the ohms of one speaker in half (speakers of equal ohms only).

Prevent siren burnout

4 ohms — Single 4-ohm speaker
8 ohms — Two speakers in series
2 ohms — Two speakers in parallel

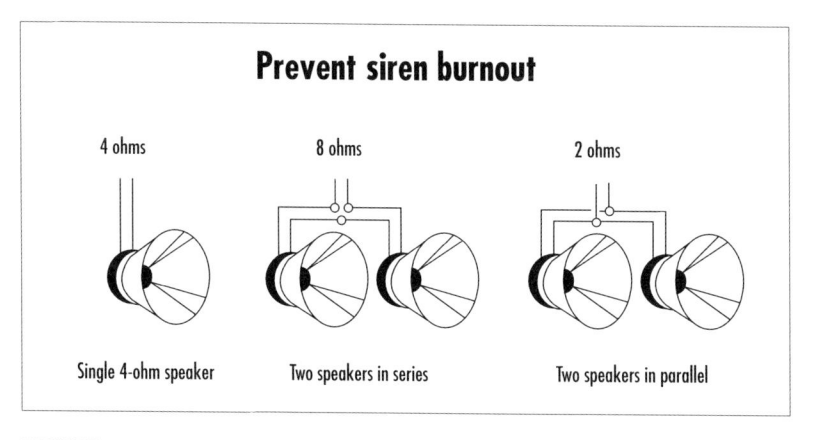

FIGURE 6–7. For optimum performance and operational longevity, connect your speakers in series/parallel combinations to assure that the total impedance matches the output of the siren driver/power amplifier.

The second option is to connect the speakers in series, providing an equivalent resistance of 8 Ω. This was computed by adding together the ohms of two speakers.

Speaker wattage, also important, must match the driver. However, when connecting speakers in parallel or series, speaker wattage can be reduced in half. This is because the current in a series speaker circuit is constant and the voltage is divided. In a parallel speaker circuit, the voltage is constant and current is divided. This results in an equivalent wattage equal to that of the driver's rated wattage.

Although speaker wattage can be reduced when connecting them in parallel and series, I prefer the wattage of these speakers to equal or exceed the driver's rated wattage. This assures greater speaker reliability over the life of the system.

SWITCH BIG LOADS

Exceeding the current specifications of a voltage source can damage sensitive electronic components. A transistor-operated relay circuit solves this problem by switching a high-current load using a low-current (trigger) voltage source.

A precut perforated board—with the foil traces already in place—is an ideal platform on which to mount an NPN transistor, diode, resistor, and a low-current relay (see the parts list in Fig. 6–8). Because the circuit requires 10 Vdc, a variable-voltage regulator (described in the section "Add Voltage Regulation to Your Alarms") also should be built on the same pc board. The regulator circuit will power more than one transistor relay circuit.

A common ground connection is needed between the panel's high-current power supply and the low-current (trigger) voltage source, which is connected to the resistor/transistor base circuit ("+ Trip voltage"). This low-current source can be a status or arming LED, for example. The relay operates when the LED turns on. The higher-current contacts of the relay then changes electrical states, switching a large load current, using the panel's power supply.

I built three transistor relay circuits when I connected a hardwire alarm panel to an eight-channel wireless receiver. The alarm panel's zone inputs were not compatible with the receiver's NO

Do-It-Yourself Circuits

Transistor relay circuit

```
                                    NO
                                    ○
                        C   ○───────
                                    ○
                  RY1               NC

    Q1        D1

                    ┤├ +10 VDC    Voltage
                                  regulator
                                  adjusted
    R1              ┤├ −10 VDC    for 10 VDC
                                  output.

                    ┤├ + Trip
                       voltage
```

Transistor circuit parts list

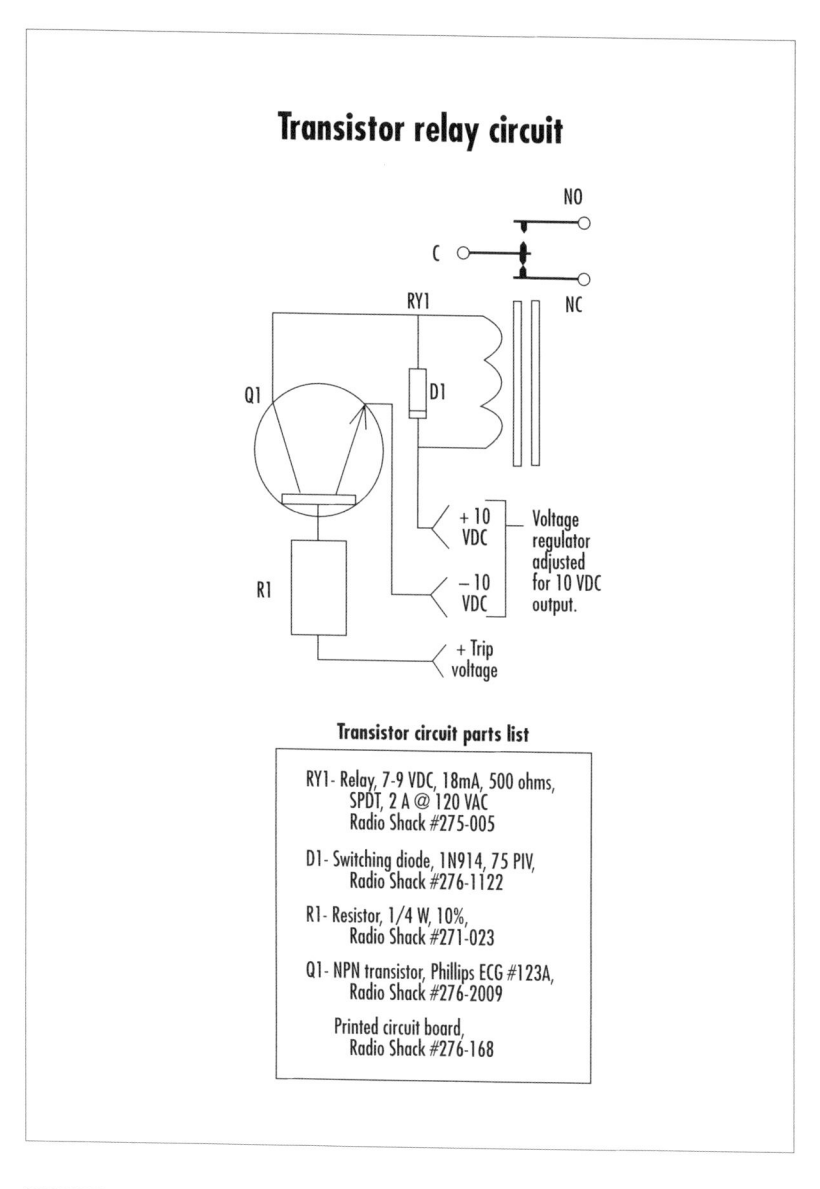

RY1- Relay, 7-9 VDC, 18mA, 500 ohms,
   SPDT, 2 A @ 120 VAC
   Radio Shack #275-005

D1- Switching diode, 1N914, 75 PIV,
   Radio Shack #276-1122

R1- Resistor, 1/4 W, 10%,
   Radio Shack #271-023

Q1- NPN transistor, Phillips ECG #123A,
   Radio Shack #276-2009

Printed circuit board,
   Radio Shack #276-168

FIGURE 6–8. The transistor relay circuit switches large loads using low-current sources.

relay outputs (the zone inputs are multiplexed on the panel's pc board). They require NC detection devices—without end-of-line resistors.

I used the low-current (4.8 Vdc @ 10 mA max.) output from a bank of eight LEDs to trigger the transistor circuit. The power supply in the alarm panel then powers the relay. Every time a wireless zone is violated, a corresponding LED lights up. This activates the transistor, turning on the transistor and relay.

CONTROL HIGH VOLTAGES WITH LED POWER

When an Oregon security dealer needed to activate a motorized gate using LED power, he designed this MOSFET (metal-oxide semiconductor field effect transistor) circuit (see Fig. 6–9). This MOSFET circuit features its own plug-in 16-Vac transformer, bridge rectifier, 2,000 1 µf filter capacitor, and an NTE67, N-channel MOSFET. To control a high-voltage device, just add a 24-Vdc relay with a

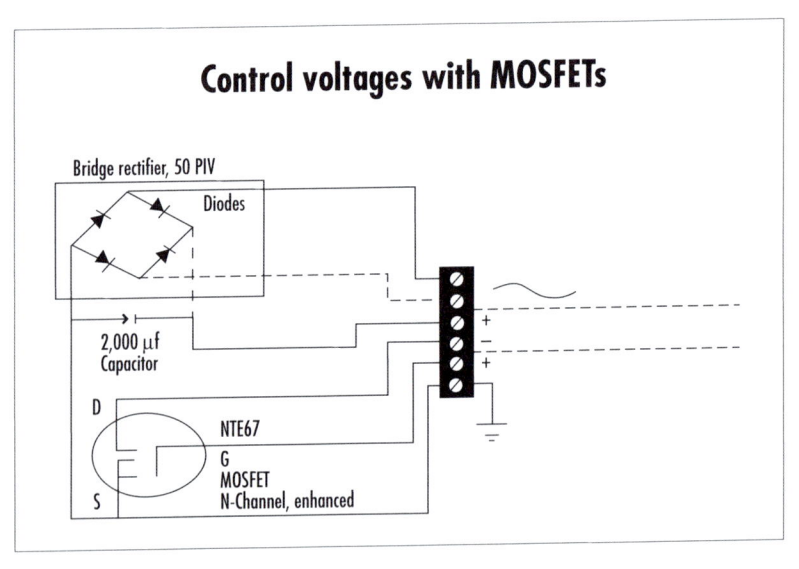

FIGURE 6–9. Control voltages with MOSFETs.

Do-It-Yourself Circuits

normally open/normally closed output. Be sure the current/voltage rating of the relay exceeds that of the high-voltage device.

BUILD THE SIREN-TO-RELAY CIRCUIT

Connecting a digital communicator to an alarm panel with only a siren (audio) output can be frustrating. I recently experienced the same problem when I tried to connect a shelf-mounted, self-contained burglar alarm system to a central station.

I prepared for the job by purchasing and programming a four-channel digital communicator. Because the panel lacked a 12-Vdc auxiliary output, I also purchased a small power supply card, a 2.5-AH rechargeable lead-acid battery, and a plug-in, low-voltage transformer.

I built a small relay circuit that connects the panel to the digital communicator (see Fig. 6–10) because audio outputs are not suitable for powering relays. The circuit consists of a 12-Vdc relay, an electrolytic capacitor, and a diode.

Convert audio sirens to relay outputs

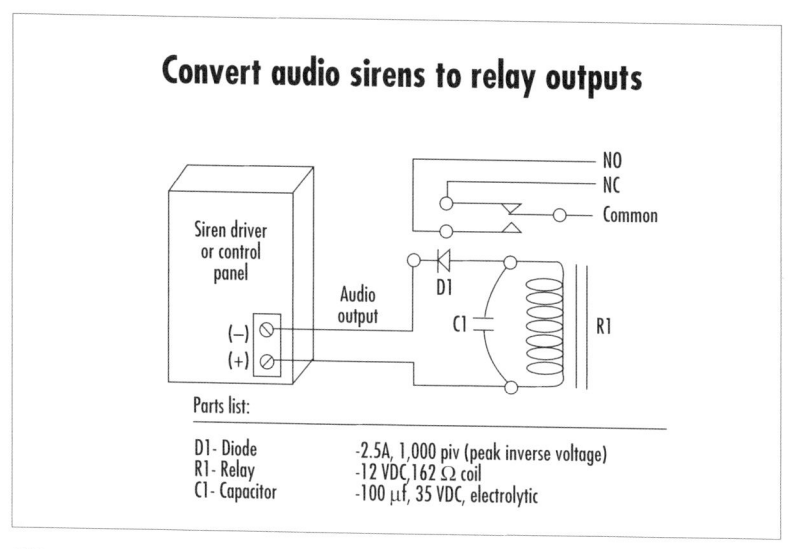

Parts list:

D1 - Diode — -2.5A, 1,000 piv (peak inverse voltage)
R1 - Relay — -12 VDC, 162 Ω coil
C1 - Capacitor — -100 µf, 35 VDC, electrolytic

FIGURE 6–10. This simple relay circuit converts an amplified audio signal into dry-contact relay outputs. Here are some possible applications: triggering digital dialers, strobe lights, power-line carrier modules, lighting up remote annunciator lamps, and activating a louder siren output.

The electrolytic capacitor (100 1 µf, 35 Vdc) is placed over the relay to filter the half-wave dc. It smooths the pulsating dc by slowing down the decay of each inverted half-cycle. This prevents the relay contacts from intermittently opening or chattering.

The alarm panel in this case already was equipped to emit a pulsating dc voltage. However, because some panels send out ac audio signals, a diode should be added. The diode assures that rectification takes place when ac audio (siren) signals are received.

This circuit was a life saver. However, keep in mind that it cannot tell the difference between a yelping, steady, or some other audio sound. However, it does convert audio to a single, dry-contact output to trigger an ancillary device.

BUILD AN EQUIPMENT LOAD TESTER

The equipment load tester (see Fig. 6–11) is a multipurpose tool. Along with a digital volt-ohm meter, it will test gel-cell alarm batteries and siren drivers under load, according to Dominion Security Systems Inc., Alexandria, VA.

Most of the parts necessary to build the load tester can be purchased from Radio Shack. The #67 lamps, however, will have to be purchased at an electronic parts store. Be sure to buy a small perforated printed-circuit board to mount the parts on.

The dealer says, to operate the tester, fasten the tester's alligator clips to the DVOM's probes. Then hold the probes to each side of a battery, for example. The two #67 bulbs will load the circuit, drawing 1.1 A. He specifies that the tester must load the battery for at least 30 sec. If the voltage drops below 12 V, replace the battery.

The load tester also can check siren drivers. Disconnect one side of the speaker from a driver and clip the load tester to the driver's output. The lamps will light up and the siren will be heard in the tester's onboard speaker if the driver is good.

BUILD AN ALARM SENSOR TESTER THAT WORKS

Determining the status of NO and NC security sensors always has been a lot more involved than appeared to be necessary. A simpler, more direct device for determining sensor states is possible.

Do-It-Yourself Circuits

Load tester

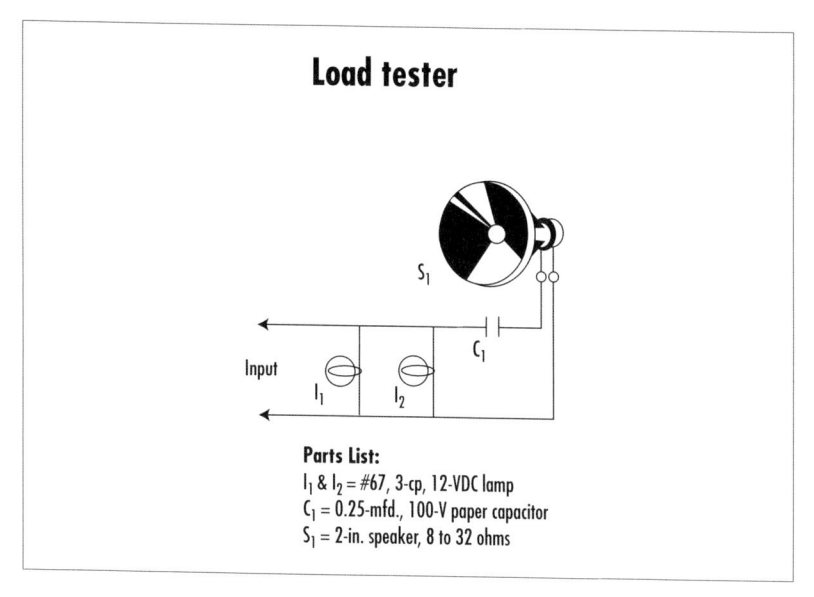

Parts List:
I_1 & I_2 = #67, 3-cp, 12-VDC lamp
C_1 = 0.25-mfd., 100-V paper capacitor
S_1 = 2-in. speaker, 8 to 32 ohms

FIGURE 6–11. The load tester checks the condition of batteries and siren drivers. To increase speaker volume, increase the value of C1.

Although an ohm meter can provide an exact look at the resistance of a sensor's output, a sensor tester is faster when only the "open" or "closed" status of a sensor is needed.

This tester is helpful during alarm system installations and service calls. For example, the open and closed condition of NC door switches; NO floor mats; form A, B, and C relays; and NO and NC motion detectors can quickly be checked. It also can be used to walk-test sensors that usually are not equipped with LED indicators.

Here is how it works. Fasten each of the clip-on probes to each side of a sensor's output circuit. Select either the NO or NC operating mode using the function switch, S2 (see Fig. 6–12). When the sensor triggers, a piezoelectric sounding device will sound, giving notice of a sensor's status. When finished, S1 disconnects the battery from the tester.

Building the alarm sensor tester

Tester shopping list

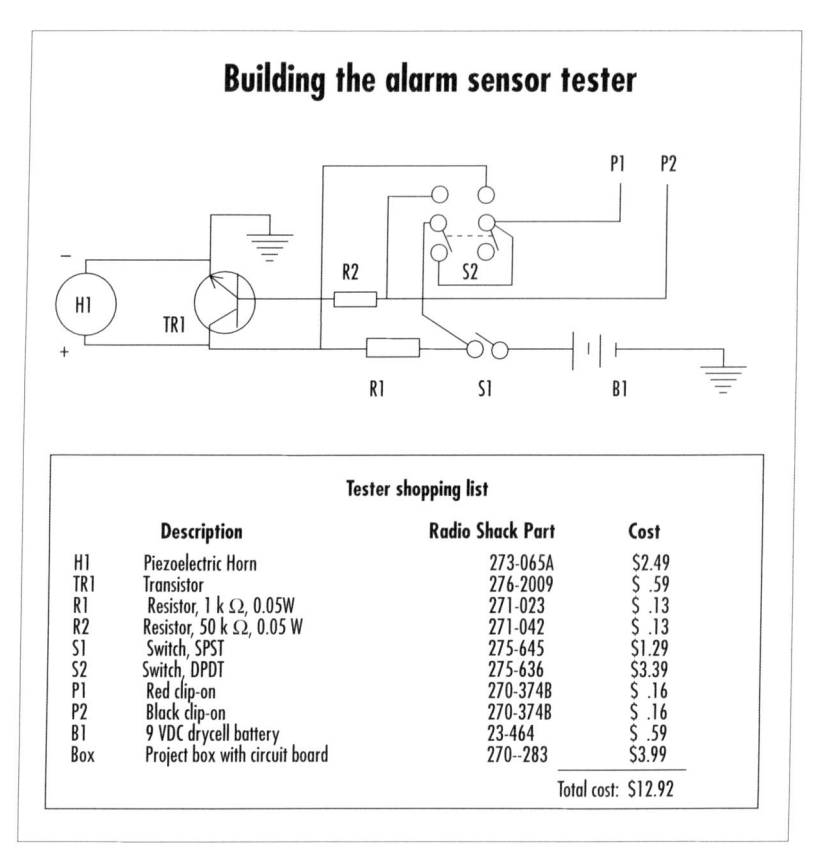

	Description	Radio Shack Part	Cost
H1	Piezoelectric Horn	273-065A	$2.49
TR1	Transistor	276-2009	$ .59
R1	Resistor, 1 k Ω, 0.05W	271-023	$ .13
R2	Resistor, 50 k Ω, 0.05 W	271-042	$ .13
S1	Switch, SPST	275-645	$1.29
S2	Switch, DPDT	275-636	$3.39
P1	Red clip-on	270-374B	$ .16
P2	Black clip-on	270-374B	$ .16
B1	9 VDC drycell battery	23-464	$ .59
Box	Project box with circuit board	270--283	$3.99

Total cost: $12.92

FIGURE 6–12. A transistor (TR1) operates a piezoelectric horn (H1) when either a closed connection is removed or an open circuit closes across P1 and P2.

The security sensor tester can be built for less than $15 using Radio Shack parts (see the schematic diagram and shopping list in Fig. 6–12). The equipment box comes with a small circuit board and cover and the 9-Vdc battery fits at the bottom of the small equipment box.

Checking sensor output is fast and easy with the sensor tester. Alarm technicians who take the time to build it will find it helpful when testing new and used sensors. Similar, commercially available circuit testers often cost as much as twice the cost of building this one.

Part 3

CCTV

One of the fastest growing segments of the security industry is CCTV (closed circuit television). Converting light into electrical signals, transmitting them, and then converting them back into light energy again is no easy task. No less simple is the task of protecting that minute electrical signal as it encounters outside electromagnetic fields in the environment as it passes through the inside conductor of a coaxial cable.

7
CCTV Kinks & Hints

UNDERSTANDING RESOLUTION

The terms *horizontal resolution*, *TV lines*, and *pixel count* can be confusing. For example, "if a camera specification states 800 horizontal pixels, that is theoretically the same as 600 TV lines," says Bob Curwin, president of Image Marketing, Schaumburg, IL.

This is because the width of a standard TV screen, established by the NTSC (National Television Standards Committee), is greater than its height by a ratio of 4:3 (also called its *aspect ratio*).

Traditionally, horizontal resolution is measured by counting the number of alternating black and white vertical lines encountered along a single horizontal reference line. The reference line itself must be equal to the height of the screen or target.

This means that the horizontal resolution of a camera is approximately three quarters of the screen's/target's full width (600 TVL/800 pixels).

Comparing one camera's horizontal and vertical pixel count to another camera's horizontal resolution is a lot like equating apples to oranges. It is possible, however, to draw a comparison by converting a camera's horizontal pixel count into a horizontal resolution value.

For example, a black/white chip camera—with no filters or intensifier—may contain 510 horizontal (H) and 492 vertical (V) pixels. To approximate this camera's horizontal resolution, multiply the horizontal pixel count (510) by .75 ($\frac{3}{4}$), for a resolution of approximately 383 TV lines.

The pixel-to-TVL formula, however, works only on black and white cameras. Also, keep in mind that adding a filter or intensifier typically degrades image resolution.

STOP THE QUAD SYSTEM JITTERS

A dealer asks, "I recently installed a quad display with dissimilar cameras. When I play back the VCR in time-lapse mode, all four of the quad camera images are unstable. Is there an easy, inexpensive way to fix this problem?"

Harvey Taylor, service manager of Gyyr Service, Anaheim, CA, says "yes." He says that Ken Madden, PhotoScan Northwest, Missoula, MT, recently called him about the same problem. According to Taylor, Madden installed one 12-Vdc and three 24-Vac cameras with a multiplexer and a time-lapse tape recorder. Madden said that, during the time-lapse playback mode, all four of the camera images in all four quadrants had a bad case of the jitters.

Taylor said that Madden discovered that, with one camera on full screen and the VCR paused in playback mode, he could stabilize all four camera scenes by changing the VCR's VLOCK (vertical lock) adjustment. He adjusted it until the vertical jitter disappeared.

For more information, contact Gyrr, 1515 South Manchester Ave., Anaheim, CA 92802, call (714) 772-1000, or fax (714) 776-6363.

LIGHTNING PROTECTION FOR CAMERAS

When lightning strikes, an immense electromagnetic field generates high surge currents in CCTV coaxial cables. This can damage camera equipment and cause fires.

"This can be prevented by installing video surge protectors at outdoor pole-mounted cameras and indoor switchers," says Roger Block, president of PolyPhaser Corp., Minden, NV. The

CCTV Kinks & Hints   89

model IS-75BB (see Fig. 7–1), manufactured by PolyPhaser, is capable of handling up to 18,000 amps of surge current on coaxial cables that carry dc power and ac signals (15 MHz max.).

Surge protectors, however, are only part of the solution. These devices require low-resistant grounds to work properly. This is because "an imperfect ground can reflect a portion of the surge back to the camera," says Block.

Grounds should be established at both ends of a coaxial cable. This is because "the cable that connects a camera to a switcher can act as a long antenna, bringing surge currents indoors," says Block.

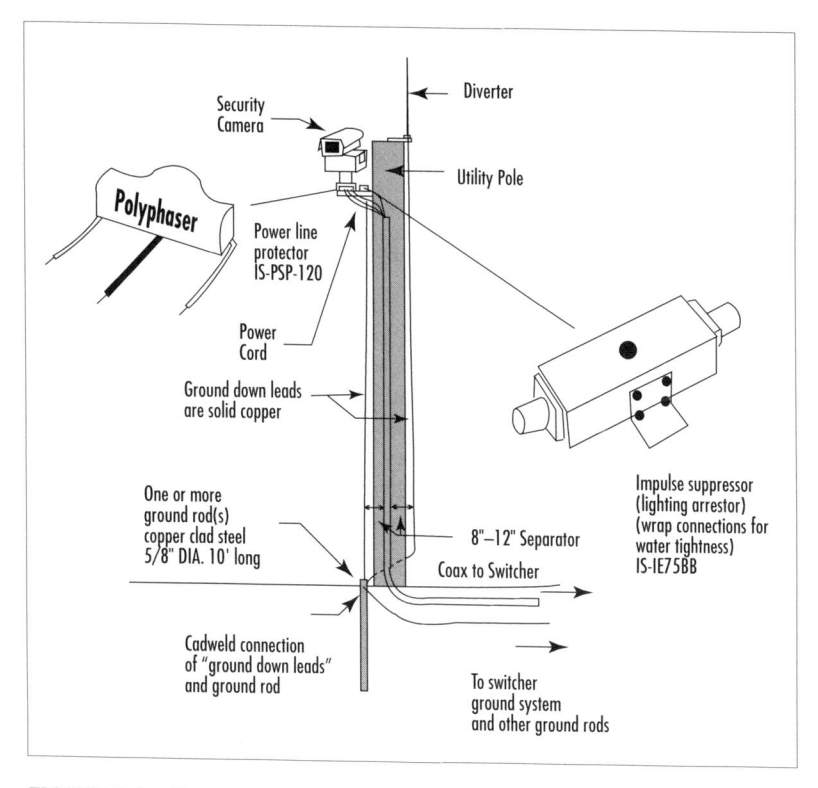

FIGURE 7–1. The model IS-75BB, manufactured by PolyPhaser Corp., Minden, NV, prevents voltage surges in CCTV coaxial cables.

Now, if a surge enters at the switcher, it is routed past the equipment to earth ground.

HOW TO PROPERLY SET UP VIDEO MONITORS

There really is a "right" way to set up a video monitor, says Burle Industries, Inc., Lancaster, PA.
Here's how to do it right:

1. Set the level of light in the room where it normally should be.
2. Adjust the "brightness" and "contrast" settings to minimum.
3. Turn the brightness to where a faint glow can be seen on the screen's raster.
4. Allow your eyes to adjust to the faint glow.
5. Turn down the brightness until it "just" disappears, then turn it up again until it "just" reappears.
6. Turn up the contrast until the video information is seen. Then adjust it until the image is comfortable to view.

Be sure not to overadjust the contrast, causing the image to smear.

KEEP SPACE BETWEEN COAXIALS AND HIGH-VOLTAGE WIRES

When security dealers install the coaxial cables used in closed-circuit television systems too close to high-voltage wires, camera images on monitor screens usually experience rippling, bending, and tearing. You can prevent this by installing coaxial cables at least 1 ft away from 110-Vac wires and 2 ft away from 220-Vac wires (see Fig. 7–2).

Because the video signals carried by CCTV coaxial cables measure only 1-Vpp, compared to the high voltages carried by electrical wires in a building, it takes little for these low-level signals to become corrupted.

For example, electrical wires that carry high-voltage alternating currents inherently emit alternating magnetic fields. These fields can surround the wires. When this alternating magnetic field comes

CCTV Kinks & Hints

Separate coaxial and high-voltage cables

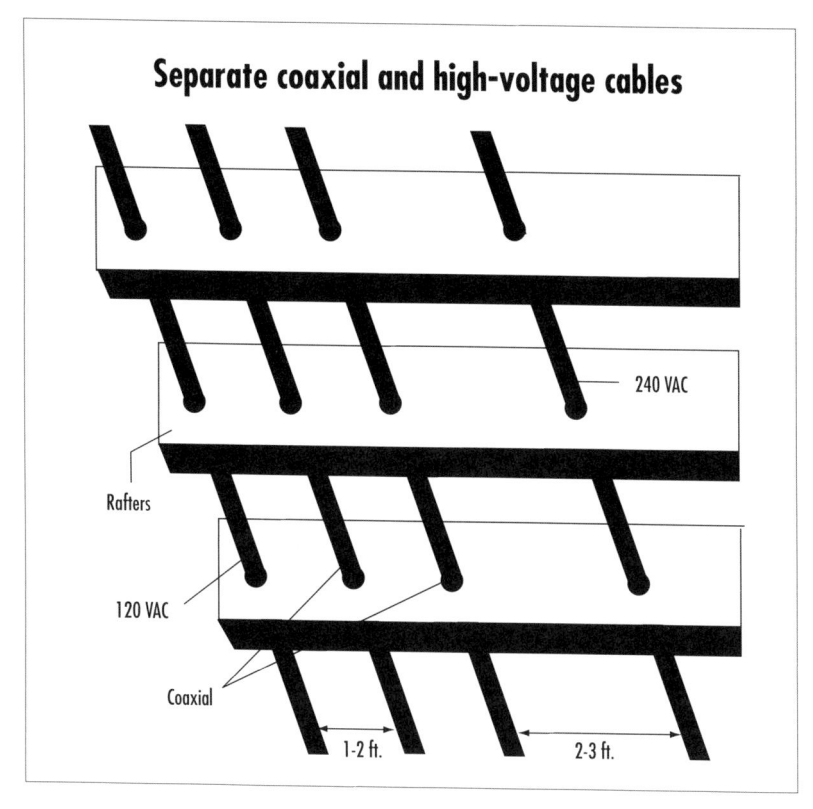

240 VAC

Rafters

120 VAC

Coaxial

1-2 ft.   2-3 ft.

FIGURE 7–2. To prevent a nearby high-voltage ac field from interfering with the video signal in a coaxial cable, maintain at least 1 ft of space for every 100 V.

in contact with a coaxial cable carrying a low-level video signal, a small ac voltage is induced in the braided shield of that coaxial cable. Because the shield is one of the signal current conductors, an IR drop develops in the shield that causes the ac voltage to become modulated onto the original video signal.

Also, an alternating magnetic field around a high-voltage wire can induce an alternating voltage in the center conductor of a coaxial cable. Although coaxial cables are shielded, not all shields are created equal. For example, a typical single-shield cable is capable of rejecting 80 percent to 96 percent of an alternating magnetic field created by the flow of electrons in a high-voltage cable.

Whether it is 80 percent or 96 percent depends on the number of conductors in the braid, how tightly they are wound together around the center conductor, and the dielectric material separating them. A precision video cable also is available that provides two shields. Although this coaxial cable costs more money to buy, it is capable of shielding the center conductor better, rejecting 100 percent of the high-voltage magnetic fields it is exposed to.

ELIMINATE HARMFUL GROUND LOOPS

Ground loops result when more than one ground path exists between two or more audio and CCTV devices. When a ground loop exists, technicians will either hear a 60-cycle hum in audio outputs or will see hum bars on CCTV monitors.

Ground loops are created when the quality of equipment ground connections and the distance between them and their earth-ground connections vary. This creates a buildup of resistance that, as ground currents flow through them, causes ground loops.

The ground currents that flow through equipment-ground and earth-ground connections (see Fig. 7–3) essentially act as a loop antenna, picking up high-voltage, 60-cycle power line interference. Because signal currents flow with ground currents through the shield of a conventional single-conductor, shielded coaxial cable, the 60-cycle interference is modulated directly onto the accompanying signal current.

Security dealers can minimize ground loops by eliminating all but one chassis-to-earth-ground connection per device. In addition, they can replace the single-conductor, shielded cable typically used to connect equipment with two-conductor, shielded cables (see Fig. 7–4).

By separating the signal path from the ground path and eliminating one of the coaxial ground connections, the chance of interference being modulated onto the low-level signal passing through the coax is greatly minimized.

CCTV Kinks & Hints

Double grounds allow ground currents

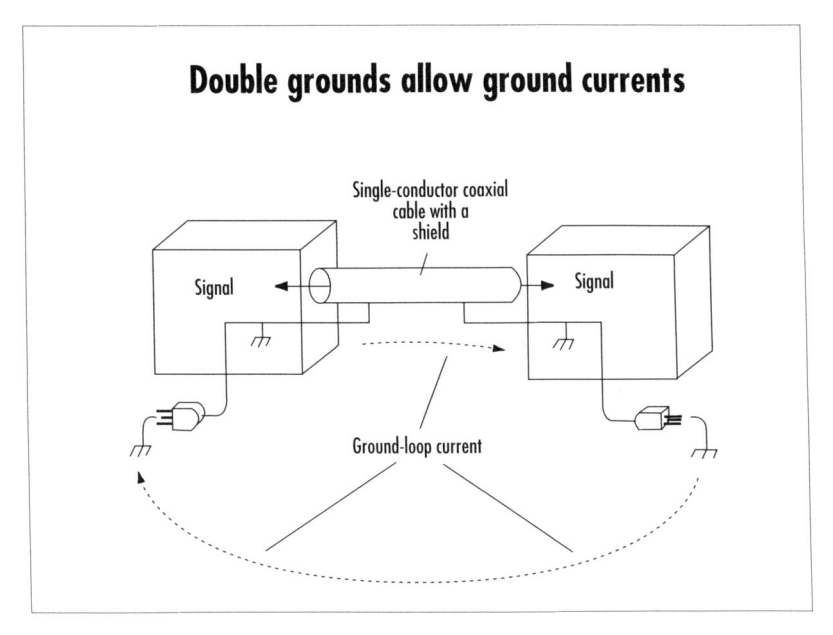

[Diagram showing two signal devices connected by single-conductor coaxial cable with a shield, both grounded, with ground-loop current flowing between them via power plugs.]

FIGURE 7–3. Ground loops occur when there is more than one ground path between devices.

Because an equipment ground is not carried between the two audio/video devices using this method, you should install balanced line transformers on each end of the new cable. Also, make sure that one side of the coaxial-cable shield on one of the chassis is broken.

For more information on eliminating ground loops, contact Yamaha Corp. of America, P.O. Box 6600, Buena Park, CA 90622, or call (815) 455-3448.

RESURRECTING TUBE CAMERAS

You can fix tube-type cameras that produce white, washed-out, and ghostly pictures, says LRC Electronics Co., Davenport, IA. All you have to do is fine tune the electron beam.

Use one ground to eliminate ground currents

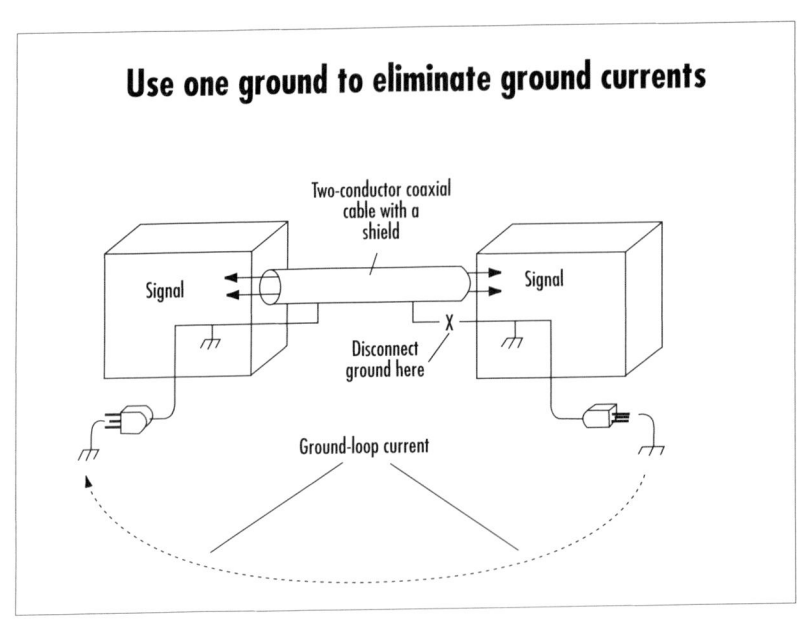

FIGURE 7-4. Two-conductor, shielded coaxial cable can eliminate ground loops.

Beam adjustments are made by inserting a small flat-blade screwdriver into the appropriate hole on a camera's cover or back plate (controls on newer models may be located inside). Turn the control one way or the other until a satisfactory image appears on the monitor. Tweaking the beam strengthens the intensity of the electron beam hitting the image target.

Another common ailment in tube-type cameras is out-of-focus images. This condition can be improved by adjusting the electronic focus control until a sharp image appears on the monitor. The focus adjustment is located near the electron beam control.

SPARE THE ROD—SPOIL THE CAMERA

Installing good grounds on both ends of outdoor coaxial video cables is essential for effective lightning protection ("Surge-Free CCTV," Kinks & Hints, February 1992). This is because it provides two pathways for lightning to travel to earth ground.

CCTV Kinks & Hints 95

Grounding two sides of a coaxial cable, however, can cause small ground-loop currents in a coaxial cable's shield. This condition usually is recognized by hum bars on the monitor and a 60-cycle hum in audio outputs. This condition is due to a potential difference between the grounds at each end of the coaxial cable.

When this happens, most technicians automatically will disconnect one of the coaxial's ground connections. Although this eliminates ground loops, it also degrades lightning protection.

Here's how to eliminate ground loops safely. Begin by driving quality copper-clad 8-ft ground rods near outdoor cameras. Use flat-stranded down-ground leads instead of round-stranded conductors. The flat leads lower the equipment-to-ground resistance by utilizing more of the flat grounding strap's surface area. This makes better use of the "skin effect," conducting more surge current to ground, ideally before damage can occur. Use solid circular conductors (#10 min.) to connect the coaxial shield on the other end to earth ground.

Then, if ground loops continue to occur, install a ground-loop network, like the model IS-IE75BB, manufactured by PolyPhaser Corp., Minden, NV. "It protects the center conductor as well as the shield," says Jim Renwick, sales manager. "However, it isolates the shield from ground, eliminating ground loops."

The IS-IE75BB should be installed to secondary grounds because indoor coaxial grounds commonly are connected to primary grounds already inside the facility. In this case, the secondary ground is installed at the outdoor camera equipment.

For more information, contact PolyPhaser Corp., 2225 Park Place, Minden, NV 89423, or call (800) 325-7170.

Part 4

Fire Alarms

The fire alarm industry is one of the most profitable businesses in the security market. Because fire alarm systems are naturally a function of life safety, they usually are required in all public buildings. The sheer number of systems installed each year is staggering, and just as amazing are the many technological advancements in fire detection.

8

Fire Alarm Installation

NFPA CLARIFIES "FAMILY"

Whether a structure meets with the NFPA definition of a one- or two-family dwelling or that of a lodging or rooming house (NFPA 101, *Life Safety Code*, Chapters 20 and 21) is important to security companies. This is because the latter requires much more by way of automatic and manual fire detection equipment than the former. A simple misunderstanding of terms when a signed contract is involved easily can mean the loss of thousands of dollars. No dealer that I know can afford that.

The NFPA, in a formal interpretation contained in the "NFPA Standards Action," December/January 1996 edition, states that *"The definition of 'family' is subject to Federal, State and local regulations..."*.

To help define its use of the term in Chapters 20 and 21, the NFPA says that a structure is a single-family dwelling when an individual or two people rent a house from someone and then sublease space to up to three other people. If the number of people to whom space is sublet increases to 4 or more, but not more than 16, the same structure becomes a lodging or rooming house. A structure is a lodging or rooming house, says NFPA, when 4 or more people, but not more than 16, rent from an owner in a facility without separate cooking accommodations.

For more information, contact National Fire Protection Association, 1 Batterymarch Park, Quincy, MA 02269, call (617) 770-3000, or fax (617) 770-0700.

USE EOL MARKERS

Not long ago it was debated in Kinks & Hints whether to place end-of-line (EOL) resistors at the end of all initiating circuits or inside the alarm control panel itself. Some dealers believe in installing them inside the panel; others do not. Still others place them on the inside of the panels using quad (four-conductor) wire with a feed and return circuit—two for the feed and two for the return circuits.

Without doubt, the safest procedure is to install EOLs at the end of each initiating circuit. However, because the same technician that installed a system will rarely return to fix it, it's not always easy to locate these EOLs.

One way to solve this problem is to have a printer make small stick-on labels with the symbol of a resistor on them (see Fig. 8–1). Dealers should place them on devices that contain an EOL. Along with EOL stickers, dealers should also maintain accurate records with the location of each EOL device. One copy should then be kept inside the panel and a second one kept in the client's file at the alarm company office.

SUBMIT THE CORRECT FIRE CERTIFICATE

Question: Why did a local AHJ (authority having jurisdiction) recently tell me that the Certificate of Compliance provided in NFPA 72, *National Fire Alarm Code*, 1993 Edition (Figure 1-7.2.1, pp. 26–29, NFPA 72) does not meet the requirements of Section 4-3.2.3.1 and 1-7.2.3 of NFPA 72, 1993 Edition, for a central station service fire alarm system I recently installed? He also said I forgot to install placards. Can you explain all of this and tell me where I went wrong?

Fire Alarm Installation

Use EOL Markers

FIGURE 8–1. Use small EOL stickers to mark the location of all EOL devices. Dealers can have stick-on labels printed by a commercial printer, like this stick-on EOL label once used by special-hazard installers.

Answer: "Currently only Underwriters Laboratories Inc., [Northbrook, IL], offers the *alarm service providers* it lists the opportunity of applying for either a UL Fire Alarm Certificate or a UL Central Station Fire Alarm Certificate" (Code Compliance, PASONA, January 1993).

According to PASONA, security dealers that are not UL certified cannot provide the NFPA-required certificate. Some dealers, instead, fill out and offer the Certificate of Compliance, Figure 1-7.2.1, pp. 26, NFPA 72, 1993. This, however, is not the certificate referred to in Sections 4-3.2.3.1 and 1-7.2.3 of NFPA 72, 1993 Edition.

In addition, placards of at least 20 sq. in. are required near the fire alarm control panel. In systems without a fire alarm panel, at least one placard must be stationed near one of the fire alarm components in the facility. These placards should contain the identify of the central station and the prime contractor by name and phone number (Sections 1-7.2.3.2.2 and 4-3.2.3.2).

For more information, contact PASONA, P.O. Box 793, Bloomfield, CT 06002, call (203) 242-4337, or fax (203) 242-2446.

SEPARATE YOUR FEED AND RETURN CIRCUITS

Question: Recently, the authority having jurisdiction on one of our larger fire alarm jobs refused to pass the system because I installed the class A return wires with the feed wires inside the same conduit. I always install my systems this way. Is there a new code I don't know anything about?

Answer: For many years, dealers were able to install both the feed and return wires in their Class A initiating circuits inside the same cable sheath or conduit. This is no longer the case.

NFPA 72, Section 3-4.2, 1993 Edition, says "All styles of Class A circuits using physical conductors shall be installed such that the outgoing and return conductors, exiting from and returning to the control unit respectively, are routed separately. The outgoing and return circuit conductors shall not be run in the same cable assembly, enclosure, or raceway."

This, of course, is to assure continuity through at least one set of wires if physical damage should occur. Whether or not this means that you have to separate these two circuits by inches, feet, or tens of feet is up to the AHJ. My advice to you and anyone else installing fire alarms is to first check with the AHJ before you install your wires.

SMOKE DETECTORS IN BEDROOMS

Question: One of our competitors is going around telling people that they must have smoke detectors installed in every bedroom when they build a new home. Isn't it true that code requires only a smoke detector on each floor, including the bedroom hallways?

Answer: Actually, your competitor is right in what it is telling people. Section 2.2.1.1.1, NFPA 72, *National Fire Alarm Code*, 1993 Edition, says that a smoke detector must be installed in each sleeping room (bedroom) in new homes. This is in addition to one on each level of the home, including outside each sleeping area (see Fig. 8–2).

Fire Alarm Installation

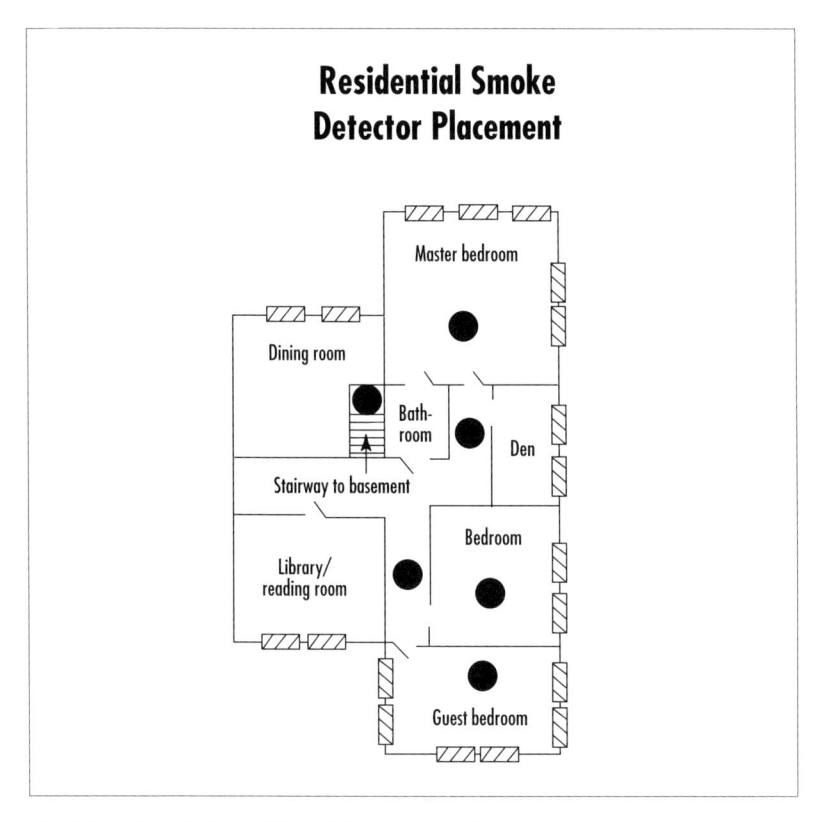

Residential Smoke Detector Placement

FIGURE 8–2. NFPA 72, 1993 Edition, requires the installation of a smoke detector on each level of a home as well as in each bedroom of a newly constructed home, like the one shown here.

Although the code is not technically specific regarding what kind of smoke detector to use, clues are given in the wording of Sections 2-2.1.1.1 and 2-2.2.1 that tell us what kind of smoke detectors to use.

Because often it is difficult to hear a self-contained smoke detector when it goes into alarm in a remote part of the house, all of the detectors in a home must be connected together so that when one goes off they all sound off. "Where there are more than one smoke detector required by Section 2-2.1, each one will be arranged so that operation of any smoke detector shall cause the alarm in all smoke detectors within the dwelling to sound" (Section 2-2.2.1).

As a whole, this section, including its Exception clause, tells us that we can use either 110-Vac, tandem-line or systems-type smoke detectors. When one of the 110-Vac, tandem-line detectors goes into alarm, the sounders inside all of the detectors will sound. Tandem-line detectors also are available with batteries that will provide backup power in case the electricity should fail.

In essence, the Exception clause at the end of Section 2-2.2.1 states that tandem-line smoke detectors are not necessary when an equivalent alarm distribution method is used. This means that dealers can employ system-type smoke detectors when a sufficient number of alarm notification appliances are installed throughout the home. Either way, a thorough test is required with all noise sources engaged to assure that the alarm is easily heard from inside each bedroom area with these doors closed.

For more information, contact National Fire Protection Association, 1 Batterymarch, Quincy, MA, phone (617) 770-3000, or e-mail at nfpa72@nfpa.org.

HOW TO COMPUTE STROBE INTENSITY (1)

Question: Recently I was asked to specify my first ADA-compliant fire alarm system in a commercial facility. I was confused about what strobe lights to install, where to mount them, and how many that ADA requires. Can you direct me to the proper agency so I get it right the first time?

Answer: Before security dealers begin installing fire alarm systems in commercial facilities, they should obtain a copy of UL1971, Underwriters Laboratories Inc., Northbrook, IL, and NFPA 72, *National Fire Alarm Code*, 1993 Edition, National Fire Protection Association (NFPA), Quincy, MA.

Regarding the type of strobe required, Section 4.28.3 of the Federal Register (28 CFR Part 36, July 26, 1991, p. 52), says that the lamp of the strobe you use must be xenon or an equivalent. The color must be clear or nominal white, it must flash between 1 and 3 times a second, and the maximum pulse duration must be 0.2 sec. with a duty life of 40 percent.

Fire Alarm Installation

Strobe specifications like these, of course, usually mean more to the engineers who design and manufacture strobes than to the security dealers who install them. Professional installers are more interested in the mechanics behind selecting the strobe they need. The intensity (candelas) of the lamp as well as how many strobe units they need in each application are more important.

"The three strobes that dealers call for the most have an intensity of 15, 75 and 110 candela," says Mike Minieri, president and CEO, TOFLAN CORP., Orlando, FL. "The 15 and 75 candela strobes are generally used in public areas, depending on the mounting height; and the 110 candela strobes is used in other areas, particularly sleeping rooms."

Underwriters Laboratories Inc., Northbrook, IL, calls for 110 candela strobes in sleeping rooms when the mounting height is more than 24 in. from the ceiling; 177 candela when the mounting height is 24 in. or less; 110 candela when the detector and signaling device are not in the same room; and 15 candela in all nonsleeping areas (Table 27.1, UL 1971).

To find the number of strobes required for a specific application, it is necessary to compare the size of each room or hallway with Tables 6-4.4.1(a) and 6-4.4.1(b) in NFPA 72 (pp. 72–96; 1993 Edition). For example, in terms of the 15-candela strobes, as specified by UL 1971, when installing wall-mounted units, only one is required in a 20 x 20 ft room, two in a 30 x 30 ft room, and four in a 40 x 40 ft room. See Table 6-4.4.1(a) for additional information on other strobe and room-size combinations.

In hallways 20 ft wide or less, which represents most of the hallways that dealers encounter, Table 6-4.4.2, NFPA 72 (1993 Edition), specifies the number of 15-candela strobes that dealers must use to properly cover the area involved.

0 to 30 ft = 1 strobe
31 to 130 ft = 2 strobes
131 to 230 ft = 3 strobes
231 to 330 ft = 4 strobes
331 to 430 ft = 5 strobes
431 to 530 ft = 6 strobes

Dealers must know some additional things before they begin installing strobe lights. Section 6-4.4.2.2, NFPA 72, 1993 Edition, says that "visible appliances shall be located no more than 15 ft from the end of the corridor with a separation no greater than 100 ft between appliances. Where there is an interruption of the concentrated viewing path, such as a fire door, an elevation change, or any other obstruction, the area shall be considered as a separate corridor."

For more information on UL 1971, contact United Laboratories Inc., 333 Pfingsten Rd., Northbrook, IL 60062, or call (847) 272-8800. For more information on NFPA 72, *National Fire Alarm Code*, 1993 Edition, contact the National Fire Prevention Association, 1 Batterymarch Park, Quincy, MA 02269.

HOW TO COMPUTE STROBE INTENSITY (2)

Question: How is it possible for the Americans with Disabilities Act (ADA) to specify a minimum light intensity of 75 candela at a distance of 50 ft and yet UL1971 specifies 15-candela strobes at a distance of 20 ft. Which criteria are we supposed to go by, the ADA or UL1971?

Answer: In practice, you should use UL1971 because it actually exceeds the ADA requirements. This is because the equivalent luminosity specified by UL1971 actually is higher than that the ADA calls for at the reduced distance specified. To fully understand this, you have to examine two or three simple algebraic formulas that define the relationship among distance, illumination and light intensity.

For example, using the formula E (illumination) = I (intensity)/D^2 (distance squared), the luminosity of a 75-candela strobe at 50 ft is 0.030 lumens:

$$E = I \text{ (candela)}/D^2 \text{ (sq. ft.)}; E = 75/50^2;$$
$$E = 75/2500; E = 0.030 \text{ lumens}$$

UL1971 assumes the distance between observer and each strobe to be no more than 20 ft. Using the same formula only with the UL1971 figures plugged into it:

$$I/D^2; E = 15/20^2; E = 15/400; E = 0.0375 \text{ lumens}$$

This clearly shows that a 15-candela strobe at 20 ft produces a luminosity of 0.0375 lumens, which is actually more than the 0.030 lumens specified by the ADA.

Fire Alarm Installation

In practice, it is possible for dealers to calculate the intensity of the strobe units they need for a given distance by using a variation of the preceding formula: $I = D^2 \times 0.0375$.

Although this formula provides the means by which dealers theoretically can calculate strobe luminosity, it does not provide the necessary requirements set forth in NFPA 72, *National Fire Code*, 1993 Edition, which must be considered before installing a fire alarm system.

For more information, refer to Kinks & Hints, *SDM*, September 1996, or contact the National Fire Protection Association, 1 Batterymarch Park, Quincy, MA 02269, call (617) 770-3000, fax (617) 770-0700, or visit its web site at http://www.wpi.edu/-fpe/nfpa.html.

ADA CLARIFICATION OF "ADJOINING ROOM"

Question: Our company designs ADA-compliant fire alarm systems for use in residential occupancies. In this application, each living area has a small powder room or water closet. Is it the intention of the ADA to regard these as separate rooms requiring their own source of illumination? Also, what is the AHJ (authority having jurisdiction) in these matters?

Answer: Roxwell Smith, SE District Manager, Cerberus Pyrotronics, Cedar Knolls, NJ (RoxellS@aol.com), says that residential occupancies are exempt from the ADA. "My interpretation of ADA is that condominiums and apartments are private homes, not public places of accommodation," says Smith. "ADA and its relationship to fire alarm specifically refers to public accommodations only."

From my own investigation on this issue, Smith is correct. Section 4.28.1, Federal Register, 28 CFR Part 36, says that compliant strobes must be installed in areas of common use. This document mentions restrooms, meeting rooms, hallways, and lobbies as examples.

Although ADA does not require strobes and other devices for disabled people in private accommodations, security dealers still cannot ignore the local AHJ. This is because, more than ever before, communities are working to develop their own disability-related mandates that may or may not be based entirely on ADA.

In this case, "the final say so . . . is in the hands of the [local] AHJ," says Kenneth Mayer, P.E., Mayer Engineering Associates Inc., Haledon, NJ (mayerk@haven.ios.com). "Although ADA is a civil rights law, [which] is not within the jurisdiction of the AHJ to enforce, it makes sense to comply with the CABO/ANSI, A117.1 standard [because many] jurisdictions use the BOCA code, which is based on CABO/ANSI, A117.1." Mayer says that BOCA also uses the CABO/ANSI standard in this regard.

For more information, contact Building Officials and Code Administration International, 4051 W. Flossmoor Rd., Country Club Hills, IL 60478, or call (708) 799-2300.

HERE'S HOW TO COMPLY WITH THE ADA

Some 43 million Americans have one or more physical or mental disabilities. This number will increase as the population as a whole grows older. The Americans with Disabilities Act will help those individuals live more comfortably by removing barriers that create inaccessibility and promote discrimination.

Title III of the act specifically prohibits discrimination in places of public accommodation and commercial, privately owned building establishments. That's where the new rules will have an effect on the security dealer or installer.

For fire alarm systems, the ADA requires modification or replacement of the strobes, smoke detectors, fire alarm pull stations, and other devices that provide visual or audible signaling, as well as the "operating mechanisms" of a system (see Fig. 8–3). In addition, the entire fire alarm system may have to be modified or changed to comply with the law.

For example, flashing alarm lights must be used with smoke detectors. All strobes and lights must be of a certain strength and flash rate.

Affected fire alarm devices include these:

- Controls and operating mechanisms.
- The highest operable part of a manual fire alarm box cannot be installed higher than 48 in. above the floor if only a forward reach is possible and no higher than 54 in. if the

Fire Alarm Installation

ADA device rules

Visual signaling on smoke detectors

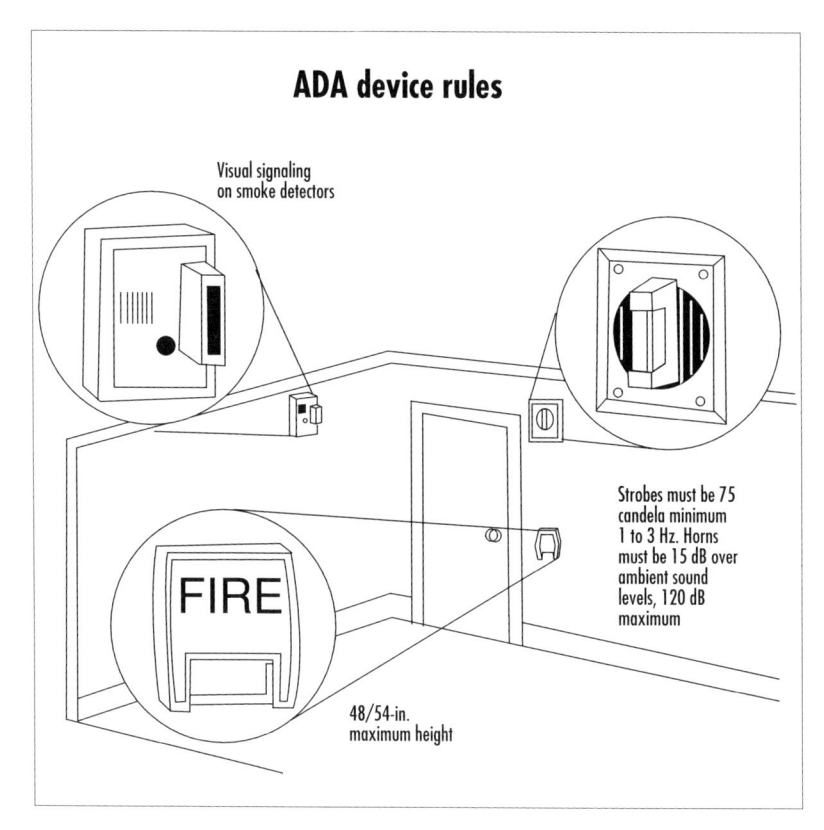

Strobes must be 75 candela minimum 1 to 3 Hz. Horns must be 15 dB over ambient sound levels, 120 dB maximum

FIRE

48/54-in. maximum height

FIGURE 8–3. The ADA has changed the mounting height of pulls, the candela output of strobe lights, and the decibel output of fire horns.

clear floor space allows a parallel approach by a person in a wheelchair.
- Visible signaling appliances. These devices must be provided in buildings and facilities in restrooms and any other general usage areas, such as meeting rooms, hallways, and lobbies.
- Audible devices. ADA rules specify that audible devices provide a sound that exceeds the prevailing sound level in the room or space by at least 15 dB or exceeds by 5 dB any maximum sound level that could have a duration of 60

sec., whichever is louder. Sound levels for alarm signals shall not exceed 120 dB.
- Visible signals. Flashing lights, strobes, or other visible indicators must be integrated into the building or facility fire alarm system. If single-station audible alarms are provided, then single-station visible alarm signals also must be included. Lamps must be a xenon strobe type or equivalent and the color must be clear or nominal white. The intensity must be a minimum of 75 candela and the flash rate must be a minimum of 1 Hz and a maximum of 3 Hz (see the sections in this chapter on "How to Compute Strobe Intensity").

The ADA requirements affect new construction, excluding non-transient-residential occupancies—apartments, condominiums, dormitories and one- and two-family dwelling units—designed and constructed for occupancy after January 26, 1993.

The new requirements also affect remodeling and alterations performed in places of public accommodation or a commercial facility after January 26, 1992.

Places of public accommodation include places of lodging, establishments serving food or drink, sales or rental establishments, service establishments, stations used for public transportation, places of public display or collection, education centers, social service establishments, and places of exercise or recreation. Commercial facilities include those locations intended for nonresidential use by a private entity and whose operations affect commerce. Included in this category are factories, warehouses, and office buildings in which employment occurs.

In addition to installation contractors, ADA will have an impact on manufacturers, according to Larry Neibauer, president and executive director of the Automatic Fire Alarm Association Inc., Lake Mary, FL.

"Power supplies, [for smoke detectors] for example, have been downsized, but now they must be able to accommodate visual indicating appliances in many instances," he said. In addition, entities such as Underwriters Laboratories, the National Fire Protection Association, and building codes officials are wrestling with enforcement and equivalency standards, he added.

Fire Alarm Installation

Other criteria that must be met:

- Manual fire pulls must be installed so that the highest portion of a pull is no more than 48 in. above floor level when only a forward reach is possible, or 54 in. if parallel approach is possible.
- Visible signals must be integrated into all fire alarm systems.
- Lamps must be xenon type, the color clear or nominal white, and the intensity at least 75 candela.
- The flash rate must be between 1 Hz and 3 Hz.
- Audible devices must provide sound level that exceeds the ambient noise in the environment by at least 15 dB.
- Sound levels cannot exceed 120 dB total.

For additional information on the ADA, dealers should call their local and national burglar and fire alarm associations, fire alarm manufacturers, and their favorite fire equipment distributor.

Dealers can receive the specific ADA rules and regulations, free of charge, by calling the Office on the Americans with Disabilities Act, Washington, DC, (202) 514-0301.

BE NICET CERTIFIED

How much do you really know about the fire alarm business? Here's a good way to find out. The National Institute for Certification in Engineering (NICET) has instituted a graduated certification program designed to test your knowledge of fire codes and standards, supervision requirements, protective signaling systems, fire/smoke detectors, and basic electronics and electricity.

The NICET program has four certification levels (levels I, II, III, and IV). Level 1 provides enrollment status to installers who are new to the fire industry. Levels II, III, and IV are certification tiers requiring work experience in the fire industry and progressively difficult written examinations.

Level II, for example, requires at least two years of work experience and a written examination. Level III requires five years, and level IV ten years, in addition to a written exam. Examinations are open-book, which means that test takers are permitted to use current NFPA code and standard publications, as well as other reference materials.

NICET certification distinguishes those who fulfill the experience requirements and pass the exam as true professionals in the fire industry.

Find out more about NICET certification. Contact NICET, 1420 King St., Alexandria, VA 22314.

WHO'S IN AUTHORITY?

The expression, "authority having jurisdiction" is familiar to dealers who install fire alarms. It is used extensively in the National Fire Prevention Association's *National Fire Alarm Code* (NFPA 72), *National Electric Code* (NFPA 70), and *Life Safety Code* (NFPA 101, 1994 Edition, pp. 101–120).

"'The authority having jurisdiction' is the organization, office, or individual responsible for approving equipment, an installation, or a procedure" (*Life Safety Code*, NFPA 101). This authority can be the fire chief of a village or town, an appointed fire inspector in a metropolitan area, or an insurance company representative. In small communities it also is common to find the building or electrical inspector in charge of fire-alarm inspections.

The "authority having jurisdiction" often has the power to stop work on a job site if it is not being performed according to code. He or she also can dictate specific types of fire equipment and their mounting locations. In some cases, the fire authority can evict the occupants in a building when the fire alarm system fails to meet NFPA standards or codes or a community's fire ordinances.

BE FIRE CODE SMART

BOCA (Building Officials and Code Administrators Intl., Inc.), SBCCI (Southern Building Code Congress Intl.), ICBO (International Conference of Building Officials), and other code-making bodies update their fire alarm codes regularly. If you're installing fire alarm systems you must make it a point to know what code changes have been made.

Fire Alarm Installation   113

A *Guide to Code Requirements*, released by the National Electrical Manufacturers Association (NEMA), provides a summary of fire equipment and NFPA 101 *Life Safety Code* requirements. Its 538 pages contain code summaries for all 50 states.

A second book, *Quality Control of Automatic Fire Detection and Alarm System Installations,* helps dealers solve false-alarm problems. It also provides important information on "establishing programs to ensure highly reliable fire detection and alarm systems," says Larry Neibauer, president and executive director of the Automatic Fire Alarm Association Inc. (AFAA).

For information on these two books, contact the AFAA, P.O. Box 951807, Lake Mary, FL 82746, or call (407) 322-6288.

INSTALLING BEAM-TYPE SMOKE DETECTORS

Beam-type smoke detectors are designed to monitor large areas using small, narrow beams of infrared (IR) light. An infrared transmitter at one end of a room emits the IR beam and a corresponding receiver, installed at the opposite end of a room, monitors it (see Fig. 8–4).

Smoke is detected when it blocks the light beam. As the density of the smoke increases, the intensity of the beam decreases. This, at some point along the way, triggers an electronic detection circuit inside the detector.

A typical beam smoke detector covers 21,000 sq. ft. Assuming that the area monitored by spot-type smoke detectors is 900 sq. ft., a single beam detector can cover the same area as 24 conventional detectors. This makes beam detectors valuable in high-ceiling applications, where it's easier to test and clean the two lenses on a beam-type detector than to test and dismantle 24 conventional smoke detectors.

The physical spacing between beam smoke transmitters and receivers should not exceed 350 ft (most cases). On smooth ceilings, beam detectors should be installed so their IR beams are 30 ft to 60 ft apart. There also should be a 1-ft to 2-ft distance between an IR beam and a ceiling as well as at least one quarter of the beam-to-beam spacing from a room's end walls and an IR beam. Further, one half of the beam-to-beam spacing should be maintained from IR beams to side walls.

Beam smoke detectors

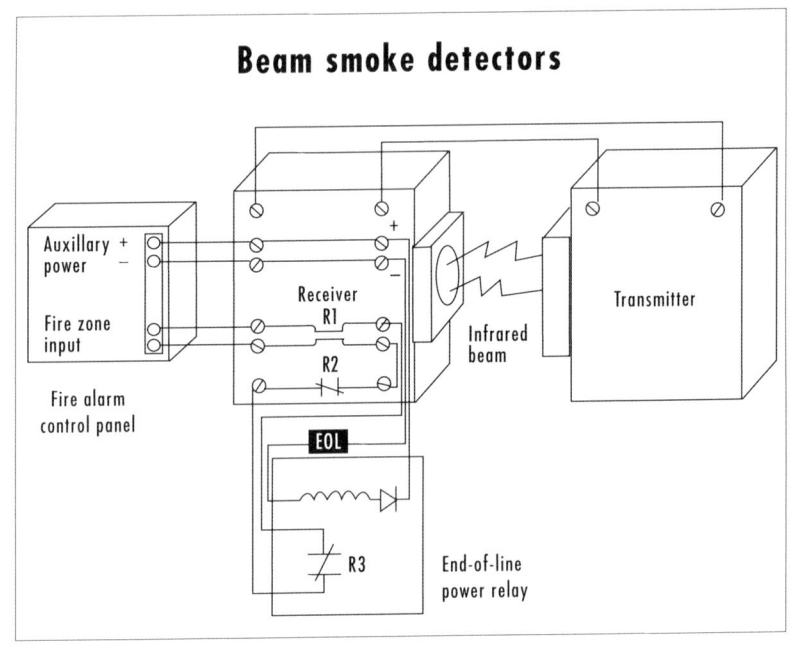

[Diagram showing: Fire alarm control panel (Auxiliary power, Fire zone input) connected to Receiver (R1, R2) with Infrared beam to Transmitter; EOL end-of-line power relay with R3]

FIGURE 8–4. Four-wire beam smoke detectors are equipped with supervision relays (R1) that activate when infrared light beams are disrupted.

For more information, contact the following manufacturers: Detection Systems Inc., (800) 289-0096; Cerberus Pyrotronics, (201) 267-1300; Edwards, Div. General Signal, (519) 376-2430; First Inertia Switch Ltd., (800) 433-6207; Gamewell Co., (508)231-1300; Hochiki America Corp., (714) 898-0795; Optex International Group, (800) 423-5669; System Sensor, (800) 736-7672.

MOUNT BEAM SMOKE DETECTORS ON CEILINGS

Frank Mioduszewski, director of support service, Detection Systems Inc., Fairport, NY, disagrees with the statement: "There also should be a 1-ft to 2-ft space between the IR (infrared) beam and the ceiling," published in "Detect Smoke with Infrared Beams" (Kinks & Hints, *SDM*, April 1994, p. 37, above).

Mioduszewski says the right way to install beam-type smoke detectors is directly on the ceiling: "we recommend that the beams be installed on the ceiling or just beneath them. This is based on the fact that smoke [in most cases] rises until it hits the ceiling. [It] then spreads across the ceiling until it hits the side walls, then it starts to bank down."

He points out that the *National Fire Alarm Code*, NFPA 72, developed and published by the National Fire Protection Association, Quincy, MA, does not list a minimum mounting distance on ceilings. "Projected beam-type smoke detectors shall normally be located with their projected beams parallel to the ceiling and in accordance with the manufacturer's documented instructions" (NFPA 72, 5-3.5.3, 1993 Edition).

In fact, beam-type smoke detectors can be installed in almost any position, as indicated in NFPA 72, 5-3.5.3, Exception No. 2: "Beams may be installed vertically or at any angle needed to afford protection of the hazard involved. (Example: Vertical beams through the open shaft area of a stairwell where there is a clear vertical space inside the handrails)."

The only time security dealers have to consider the placement of a beam-type smoke detector, according to Mioduszewski, is when the infrared (IR) light beam passes close to the upper corner of a room. "The exception to this is when the projected beam path is in close proximity to (and parallel [with]) a wall/ceiling corner. In this case, the projected beam units should be mounted no closer than 4 in. away from the corner," says Mioduszewski.

This is of special concern because the air in a room usually stratifies as it fills with smoke. The smoke-laden air then displaces the clean air at the center of the ceiling, pressing the clean air toward the upper corners of the room. This poses a serious problem because the beam then will "see" only clean air for a relatively long period of time.

For more information, contact the National Fire Protection Association, 1 Batterymarch Park, Quincy, MA 02269, or call (617) 770-3000.

WHEN AND WHERE TO INSTALL DUCT DETECTORS

Knowing how to install duct-type smoke detectors in ventilation systems is important ("Installing Duct Detectors Step by Step,"

SDM, March 1993, p. 91). Moreover, knowing when and where to install them in ventilation systems is even more important. The wrong choice can mean losing a bid unnecessarily or installing duct smoke detectors incorrectly.

Jim Vydra, president of Pro-Tech Systems Inc., Springfield, MO, says that the National Fire Protection Association's standard on HVAC systems, NFPA 90, explains when and how to install them.

He says that duct detectors are required to automatically shut down air handling systems that circulate in excess of 2,000 cfm (cubit feet/minute). "They should be installed in the supply duct downstream of the filters and in the return stream on each floor at the point of entry into the common return," says Vydra.

There are two exceptions. The first concerns a duct smoke detector in the return air stream. Vydra says that this detector can be eliminated in air handling systems that circulate less than 15,000 cfm.

Duct detectors also can be eliminated in ventilation systems that circulate less than 15,000 cfm when the ventilation system fully serves only the space in which it is installed and this space already contains an open-area smoke detection system.

NFPA 72 also requires that duct detectors be installed at least six duct widths downstream from duct openings, deflection plates, bends, and branch connections, says Vydra. An exception to this is when it is physically impossible to observe this "six-wide rule." Detectors installed closer than six duct widths must still be capable of detecting smoke.

For more information, contact the National Fire Protection Association, 1 Batterymarch Park, Quincy, MA 02269, call (617) 770-3000, or fax (617) 770-0700.

UPGRADE OLD FIRE PULLS WITH SINGLE-WIRE HOOKUPS

Today's initiating devices comply with NFPA (National Fire Protection Association) and UL (Underwriters Laboratories) standards and codes because they are equipped with either two screw terminal connections, to which incoming and outgoing wires connect, or two sets of connecting wires.

Vintage models, on the other hand, have only one set of connecting wires. The problem with this design is that, when the connection

Fire Alarm Installation 117

between the device and the initiating circuit is broken, there is no supervision/trouble indication at the fire alarm control panel.

To improve the quality of the connection between a vintage initiating device and its initiating circuit, security dealers should either keep the lead wires as short as possible or use a screw-type terminal strip (see Fig. 8–5) to connect them to the initiating circuit (Source: *The Moore-Wilson Signaling Report*, September–October 1993, p. 5).

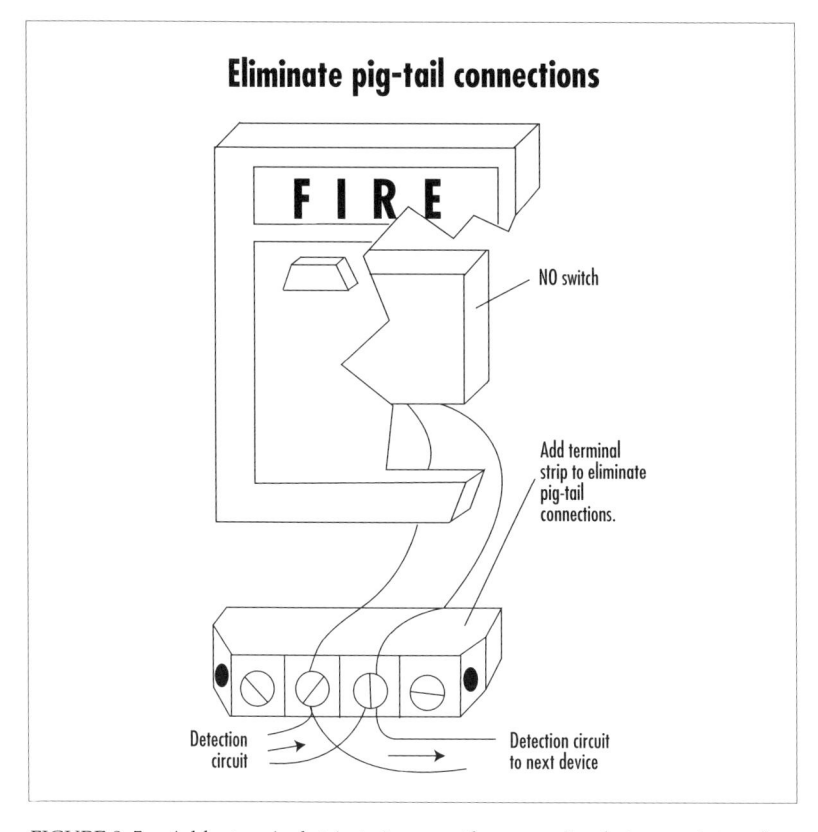

FIGURE 8–5. Add a terminal strip to improve the connection between vintage fire alarm initiating devices, equipped with only one set of lead wires, and an initiating circuit.

BACK UP YOUR ELECTRIC SMOKE DETECTORS

How many deaths does it take until smoke detector manufacturers, security dealers, and those who inspect or specify fire protection equipment realize that, by themselves, single-station battery-only and electric-only smoke detectors do not always offer enough protection? This is especially true in multiple-family apartments where single-station smoke detectors are required by the National Fire Protection Association (NFPA), Quincy, MA (NFPA 101, *Life Safety Code*), and various building-code organizations.

No doubt, building contractors, electricians, and building owners that install these single-station smoke detectors intend for their customers to use them correctly. However, tenants regularly remove smoke detector batteries and switch off power breakers when their smoke detectors sound false alarms from something as trite as cooking smoke.

Unfortunately, when these tenants forget to replace the batteries or switch the power back on to the smoke detectors, they place their families' lives in jeopardy. In many cases, history shows, these tenants often pay for this mistake with their lives or the lives of a family member.

Within the past three years, the three most used building-code-making bodies in the United States (see Fig. 8–6) have revised their building codes so they require the inclusion of battery backup in all electric-powered, single-station smoke detectors installed in multiple-family residences.

"As states and cities adopt these new codes, this will become the requirement throughout the country," says Larry Neibauer, president and executive director of the Automatic Fire Alarm Association, Inc. "This will affect states in almost every part of the country because these three model building codes are usually picked up or modified by most states and municipalities within three to five years." Many municipalities, counties, and other government agencies already have done so.

At least five states already have embraced this requirement by adopting one of the three newly revised codes: Utah, Connecticut, Virginia, Kentucky, and Maryland. Michigan and New Jersey are now considering amending their building codes to include the new smoke detector ruling.

Fire Alarm Installation

Here's who requires battery backup

Code-making bodies that require battery backup

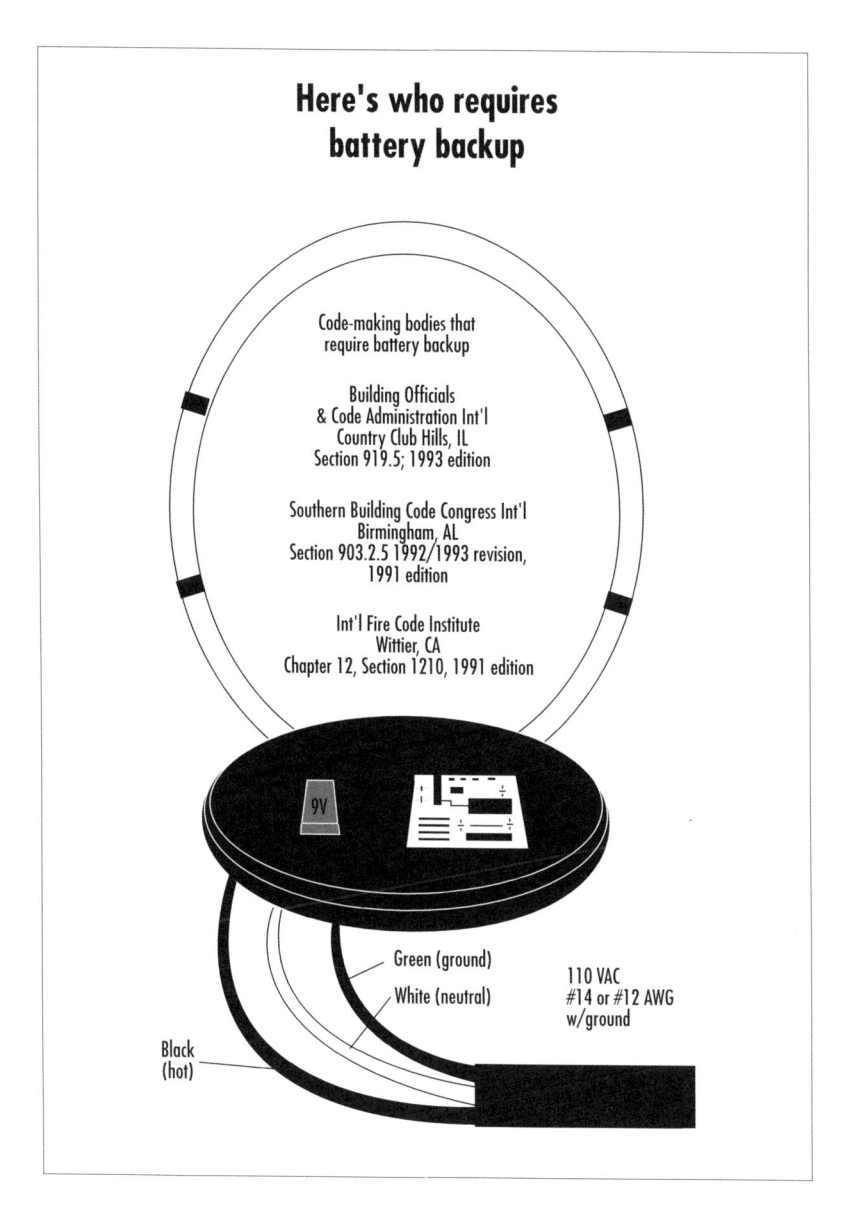

Building Officials
& Code Administration Int'l
Country Club Hills, IL
Section 919.5; 1993 edition

Southern Building Code Congress Int'l
Birmingham, AL
Section 903.2.5 1992/1993 revision,
1991 edition

Int'l Fire Code Institute
Wittier, CA
Chapter 12, Section 1210, 1991 edition

9V

Green (ground)
White (neutral)
Black (hot)

110 VAC
#14 or #12 AWG
w/ground

FIGURE 8–6. These building-code-making-bodies already have included battery-backed-up, high-voltage smoke detectors in their building codes.

For more information, contact any of the three building-code-making bodies in Fig. 8–6 or contact your local building or fire department.

CHECK FIBER LINKS ON PHONE LINES

Burglar and/or fire alarm systems that digitally report over fiber-optic-converted cables may not function properly during power outages. According to *The Moore-Wilson Signaling Report*, published by Focus Publishing Enterprises, Bloomfield, CT, the standby-power capability of fiberoptic multiplexers that convert analog signals into digital may not be up to NFPA 72, 1993 standards.

"The alarm trade associations have recently discovered two cases where the fiber-optic multiplexer installed by the telephone company at the protected premises either had no standby power at all or only eight hours of standby power" (*The Moore-Wilson Report*, March–April 1994, p. 1).

Although it's not security installers' responsibility to provide proper standby power to fiberoptic multiplexers, it is the installers' responsibility to ascertain that the listed-and-approved fire alarm and burglar alarm panels used will report up to 24 hours after power is lost, required in central station and proprietary systems, and up to 60 hours in remote-station systems (NFPA 72, 1993).

"If you suspect that the telephone company has converted your telephone system to optical-fiber cable, you should contact the telephone company to [make] certain how much standby power has been provided for the optical fiber multiplexers" (*The Moore-Wilson Report*, March–April 1994, p. 1).

NEW SIA COMMUNICATION STANDARD

The Security Industry Association's (SIA) Digital Communication Subcommittee has published a standard on common digital communicator formats. The document discusses the technical aspects of FSK, DTMF, and several tone-burst communication formats.

Get your copy of the "Generic Digital Communication Standard" by contacting SIA, 635 Slaters Ln., Suit 110, Alexandria, VA 22314, call (703) 683-2075, or fax (703) 683-2469.

Fire Alarm Installation   121

TERMINATE WIRES PROPERLY

Fire detection equipment today must have four separate screw terminals or lead wires for each in and out wire. "Duplicate terminals or leads, or their equivalent, shall be provided on each initiating device for the express purpose of connecting into the fire alarm system to provide supervision of the connections" (NFPA 72, 5-1.4.1, 1993 Editions).

This arrangement is to prevent disconnection of the feed and return wires from a single screw terminal or lead wire on an initiating device without some kind of trouble or supervisory signal occurring. In the past this was a problem, especially when pig tailing the feed and return wires to a single set of lead wires.

Use of duplicate terminals and lead wires assures that both feed and return wires are broken at the device and that individual connections must be made to the in and out screws or lead wires on an initiating or signaling device.

This is important because, if a security dealer fails to terminate these feed and return wires correctly, an initiating device can become disconnected from the circuit without initiating a trouble/supervisory signal.

The only exception to the duplicate terminal/lead wire rule is found in Addendum A of NFPA 72. Here, connection of the feed and return wires in an initiating circuit is shown to a single screw terminal. However, close examination will reveal that the screw is equipped with a teeter plate. This connection method assures that, when the connection between an initiating device and the feed and return wires is made, a trouble/supervisory signal will occur.

GIVE SUPERVISORY SIGNALS DISTINCT SOUNDS

Supervisory devices, such as sprinkler tamper switches, must produce a sound that is distinctly different from a general system-trouble sound. Otherwise, a visual method of discerning the difference between them must be provided, says the Professional Alarm Service Organizations of North America Inc. (PASONA), Bloomfield, CT.

In the past, installers connected sprinkler tamper switches in series with the end-of-line resistor after a flow or alarm pressure switch. Although this gave a general indication when the main water valve was turned (NFPA 72, 3-8.7.5, 1993 Edition), it also resulted in the same signal as the general system-trouble sound. To solve this problem, install supervisory devices on their own initiating zones.

POWER FIRE PANELS AHEAD OF THE MAINS

Nick Markowitz, owner of Markowitz Electric Protection, Verona, PA, says that security dealers should power their fire alarm control panels ahead of the main breaker(s) in a service entrance panel. "Power to a fire alarm panel should be on at all times and the circuit dedicated solely to the fire alarm," says Markowitz. He says that most people do not realize how many fire alarms have failed because the power was deliberately or accidentally disconnected.

To connect a 120-Vac fire alarm panel ahead of the main breakers in a service panel, connect the proper gauge wire (usually #14 or #12) from the input of one main breaker and the neutral buss to the appropriate terminals/wires in a 220-Vac disconnect box (see Fig. 8–7). Markowitz says to use rigid conduit to run the wires from the service box to the disconnect box.

To assure proper disconnection at the disconnect box, run the hot wire to the input side of a 220-Vac breaker in the disconnect box. Then run the same gauge wire from the hot and neutral terminals of the fire alarm panel to the appropriate breaker output. To do this, use flexible or EMT conduit.

Technicians who are not electricians should have a qualified electrical contractor perform this work. For more information, refer to Section 230-82, Exception 5, of NFPA 70, *National Electric Code*, or contact the National Fire Protection Association, P.O. Box 9101, Quincy, MA 02269, or phone (800) 344-3555.

HOW TO CONNECT 25/70V SPEAKERS TO A FIRE ALARM

A dealer asks how to add a speaker to a fire alarm voice evacuation system power amplifier with a constant line output of 70.7 Vac. First, get a step-down, audio impedance-matching transformer. A step-down transformer converts higher audio input voltages to a

Fire Alarm Installation 123

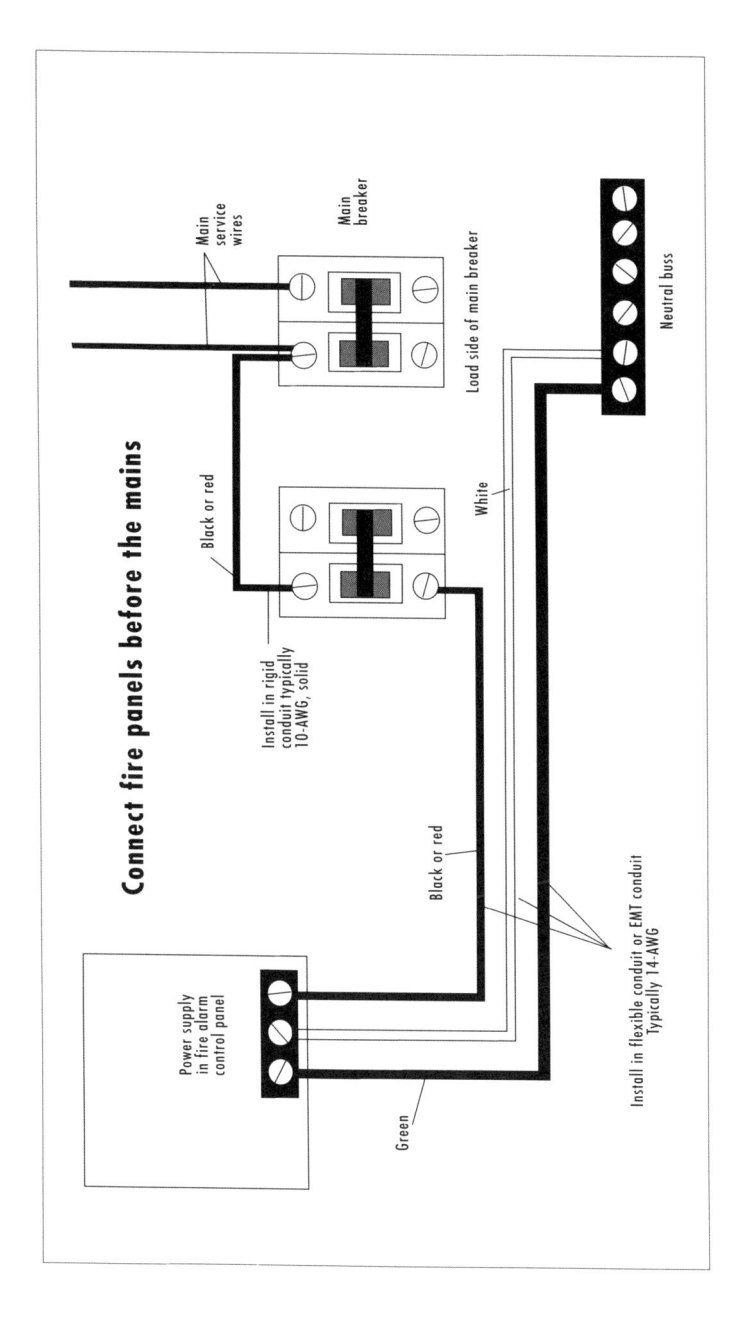

FIGURE 8–7. Nick Markowitz of Markowitz Electric Protection, Verona PA, illustrates how to provide 24-hour electric power to a 120-Vac fire alarm control panel.

lower voltage value. In this case, the maximum 25/70-Vac output from the amplifier reduces to a level acceptable to a common, ordinary loudspeaker.

Most impedance-matching transformers are universal: they come with a variety of input wattages and output ohmages from which to choose. This variety of inputs and outputs enables you to effectively match the speaker wattage and ohmage to the amplifier's constant 25/70-Vac line voltage.

For example, to add an 8-Ω, 5-W (watt) speaker to an existing 25/70-Vac line, solder one end of a two-conductor cable to the common and 8-Ω output on the transformer and the other end of the cable to the speaker.

Then solder one end of a second two-conductor cable across the common and 5-W input on the transformer and connect the other end to the amplifier's line output in parallel. Also, always consult the manufacturer's instructions before connecting to an amplifier output.

USE THE RIGHT STRANDED CONDUCTOR

An alarm installer recently asked, "In the past, I've used a 16-AWG, single-conductor wire and galvanized, thin wall conduit when installing commercial fire alarm systems. Why are we now restricted to using only 14- and 12-AWG stranded conductors?"

The *National Electric Code*, NFPA 70, Sections 760-16(c) and 760-51(a), 1993 Edition, stipulates the use of solid or bunch-tinned conductors or a conductor that has a maximum of 7 strands in an 18- or 16-AWG conductor and 19 strands in a 14- or 12-AWG conductor.

The strands in a bunch-tinned conductor, for example, are soldered together in the final phase of manufacturing by drawing them through a solder bath. This assures that the majority of strands are inserted under the screw terminals on a fire alarm device. You can use 18- or 16-AWG conductors, but they must have a maximum of seven strands.

The reason for this is simply that the more strands there are in a conductor, the smaller each of the strands has to be. This, according to the NFPA, increases the chance that a conductor will fray, possibly resulting in a noncontact situation between the majority of

Fire Alarm Installation

strands and a screw terminal. Therefore, because 18- and 16-AWG conductors have only 6 strands, 14- and 12-AWG conductors must be used because they meet NEC's 19-strand requirement.

All of this has changed, however, with the printing of the 1996 Edition of NFPA 70. Sections 760-16(c) and -51(a) will be omitted entirely. This is because the issue behind "solid or bunch-tinned (bonded) stranded copper" recently was determined to be unimportant by NFPA, as the likelihood of a terminal screw contacting a single strand is rare to none.

For more information, contact the National Fire Protection Association, Batterymarch Park, Quincy, MA, or call (800) 344-3555.

FILE YOUR FIRE ALARM SYSTEMS

Always submit a blueprint or floor plan with the "local authority having jurisdiction" before you add on, relocate, or install fire alarm equipment. To do otherwise may subject your firm to stiff fines, penalties, or expensive civil lawsuits.

One potential problem is asbestos in older facilities. Not only is it a threat to your unsuspecting employees, who may drill through asbestos-sealed surfaces, it also can release asbestos into the atmosphere—endangering the health of those who work or visit there.

City building departments and fire departments often have ordinances or laws that require a blueprint or floor plan—before work begins. The company that does not comply will eventually encounter legal problems.

To avoid this, check with the local building department or fire department *before* the installation begins. If the facility is a public building, also check with the state fire marshal's office, which may require you to submit a floor plan and written plan before you begin working.

USE PLC TECHNOLOGY TO COMPLY WITH THE ADA

The Americans with Disabilities Act, established in 1991, guarantees equal treatment and opportunity to disabled people. After January 26, 1993, existing buildings that serve the public must comply with the ADA when structural alterations are made.

The ADA-specified, 75-candela visual appliances—at a minimum—must be installed in restrooms, lobbies, meeting rooms, hallways, sleeping rooms, and other public areas in "places of lodging, food service, entertainment, public gathering, recreation, education, service establishments, and sales or rental establishments," says Wayne Carson, consulting fire protection engineer and chairman of NFPA Technical Committee on Health Care Occupancies (*NFPA's Fire News*, December–January 1993).

Compliance with the ADA can impose a large expense on your commercial and institutional customers. This is because the act requires higher-powered horns and xenon (or equivalent), 75-candela strobes with a flash rate of 1 Hz to 3 Hz.

Dealers can eliminate some of the labor associated with adding components to their systems by installing the PLC receiver, model RX-RCM, and transmitter, model TX-Bug2 (NO), manufactured by Macro Media Systems Inc., Merrimack, NH.

"Transmission of signals through the power line is relatively simple and is the basis of common, inexpensive remote light control systems" (Americans with Disabilities Act, Part III, Federal Register, Department of Justice, 28 CFR Part 36).

PLC technology, although not acknowledged at this time by NFPA as a valid signal transmission media, is accepted by the ADA. "Activation [of an audiovisual device] by a building alarm system can either be accomplished by a separate [hardwired] circuit activating an auditory alarm . . . or by a signal transmitted through the ordinary 110-volt outlet" (ADA, Federal Register, 28 CFR Part 36).

Hotel and motel rooms, for example, must be equipped with smoke (110-Vac or battery) detectors that do not connect to the building's fire alarm system. The ADA requires that a visual appliance (meeting its specifications) be either hardwired or connected through the 110-Vac power lines to a premises alarm system.

PLC technology should be used only when installing the additional components required by the ADA that are not covered by NFPA or the authority having jurisdiction (AHJ). The remainder of a system must comply with NFPA 101, NFPA 70, and the AHJ.

The RX-RCM and TX-Bug2 (NO) use the high-voltage wiring in buildings to turn on and off high-powered strobe lights and horns. This is done by transmitting on/off signals through the power lines. When installing these PLC devices, a house code must be selected

Fire Alarm Installation

using a bank of DIP switches, found on both the transmitter and receiver.

Connection to the high-voltage power lines is made using wire connectors. Because the lead-in wires on both units are 16 AWG, do not use too large a wire connector. This will prevent the wires from pulling out of the connectors, severing power with the PLC device.

The transmitter provides a SPDT, dry-contact relay output to switch on or off a horn or visual device. Whether it supplies power to the device or merely a tandem-line signal depends entirely on the application and the horn or device used.

For more information, contact Macro Media Systems Inc., P.O. Box 973, Merrimack, NH 03054, or call (603) 424-2674.

PROTECT PLUG-IN TRANSFORMERS

Fire-only systems usually use the direct-connection method to connect the primary power supply (110 or 220 Vac) to a fire alarm control panel. With the advent of combination fire and burglary alarm control panels, however, dealers are using plug-in, step-down transformers.

Although plug-in transformers work fine most of the time, they can be damaged inadvertently or removed intentionally. In addition, when they are installed without any kind of protection, they do not meet with National Fire Protection Association codes and standards.

One way to secure a plug-in transformer, according to the Professional Alarm Services Organization of North America, Inc., Bloomfield, CT, is to install it in a secure metal cabinet (see Fig. 8–8). Then secure the hinged lid with either self-tapping screws or a cam lock.

For more information, refer to NFPA's *National Fire Alarm Code*, NFPA 71, 1989, 2-2.4.1, NFPA 72, 1990, 5-4.2, or contact NFPA, 1 Batterymarch Park, Quincy, MA 02269, or call (617) 770-3000.

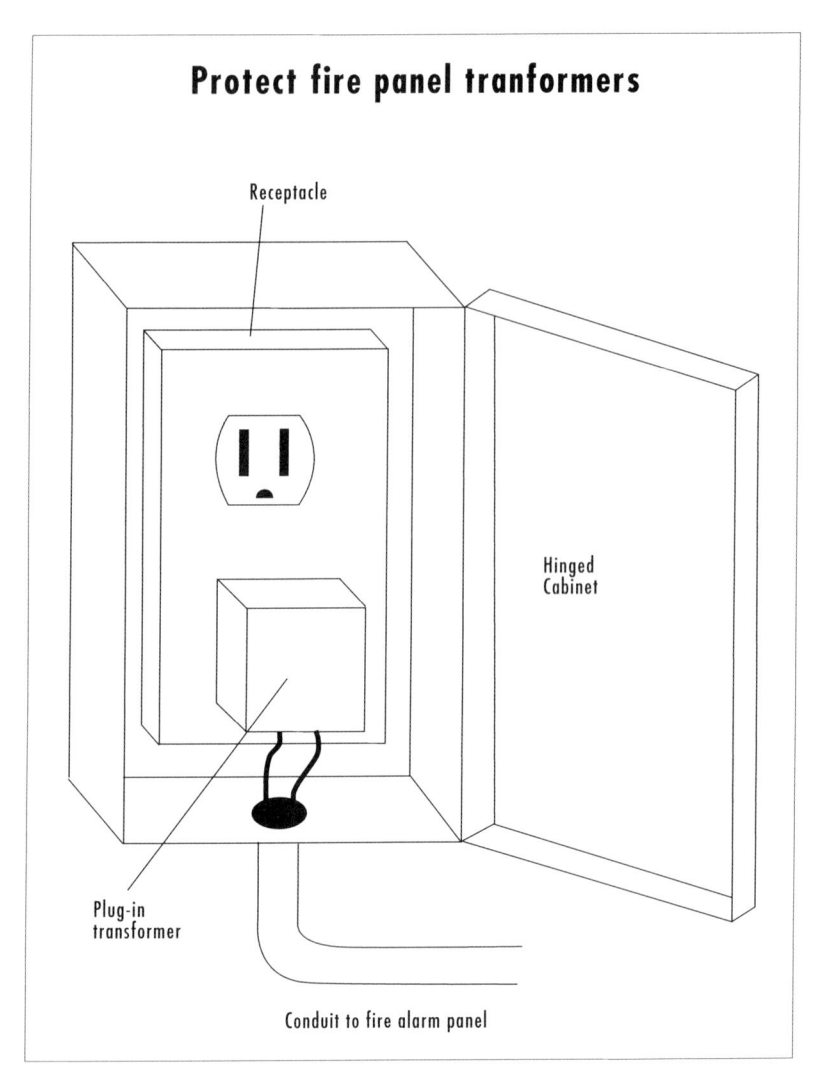

FIGURE 8–8. Protect plug-in transformers that power fire alarm systems by installing them in a metal box with a securable lid.

9

Servicing Fire Alarms

UPDATING OLD FIRE ALARM SYSTEMS

Question: What's the best solution when faced with replacing an old series, gong-type fire alarm system in an old installation where installing new wires is not an option?

Answer: Nick Markowitz, president, Markowitz Electric Protection, Verona, PA, recently experienced the same problem in the Pittsburgh area. He solved it by removing the gongs and converting the series gong circuit to a conventional Class B, notification appliance circuit. He then added an EOL to the end of each Class B circuit for supervision.

Changing an old series-gong circuit to a conventional Class B circuit is not as difficult as it may sound. This is especially true in installations where the wires are installed inside conduit and the gongs are mounted to metal boxes. Conversion in most cases also will require the changeout or modification of the old coded fire pull stations.

There are two courses of action to take in this case. The first one requires engineering special back plates—made to fit the oversized fire pull holes. This usually will include cutting a smaller hole in the center of the plate to accommodate the new models. The alternative, as Markowitz discovered, is to modify the old pulls so they work with the conventional fire alarm system. This was accomplished in Markowitz's case by removing the code wheels inside the old pulls and replacing them with solid Lexan disks (see Fig. 9–1).

Gong-type alarm conversion

Gongs

EOL

Horns

Old gong-type circuit

New converted Class B notification appliance circuit

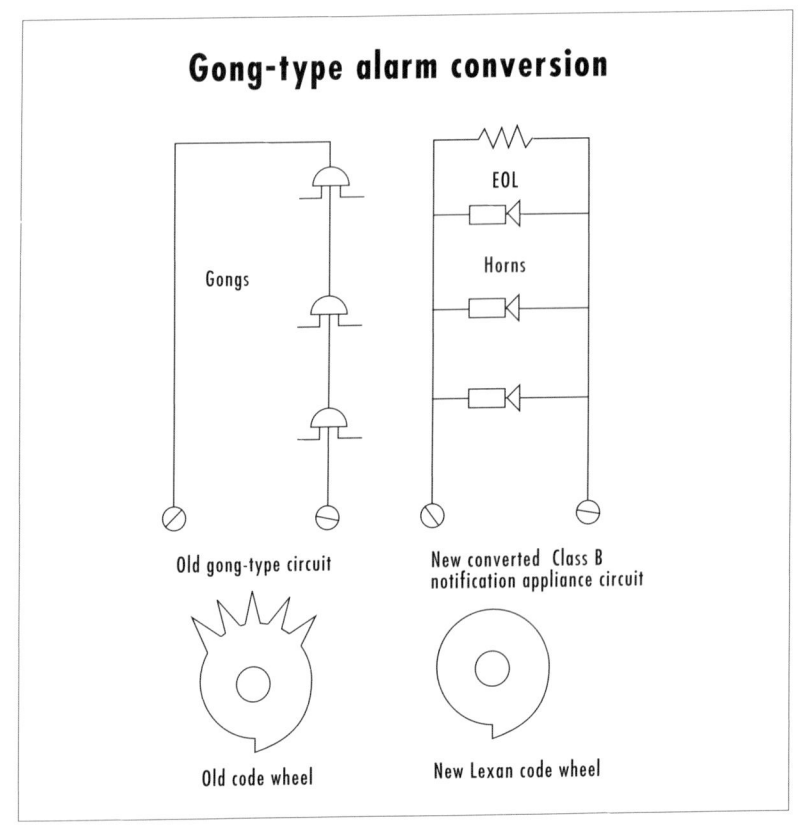

Old code wheel

New Lexan code wheel

FIGURE 9–1. When pulling new wire is out of the question, Nick Markowitz, Verona, PA, converts his old series-gong fire alarm systems to conventional specifications.

SOLVING FIRE ALARM GROUND FAULT PROBLEMS

Question: Troubleshooting ground faults in a fire alarm system is terribly time consuming. Is there an easier way to find this type of problem other than an ohm/volt meter and trial and error?

Answer: Here's a method that allegedly will quickly point you in the right direction. Keep in mind that I have not verified the authenticity of this method or whether it's safe to use.

"Tradition dictates that you start [by] removing each pair of initiating circuits from the [fire alarm control] panel until the ground fault clears. [T]hen you start looking for the fault within that zone," says Allan C.W. Hunt, Installation Supervisor, Sentry Alarms Inc., Abbotsford, BC. "Here's a shortcut that works almost every time, unless the short to ground is very high [in] resistance."

Hunt says (see Fig. 9–2) to

- Notify the central station and/or fire department that you are going to work on the system.
- Place the fire department disconnect switch (if applicable) in the "off" position.
- Connect a short length of wire to the panel's earth-ground connection and quickly touch the other end to the positive (+) side of the standby battery.
- If the system fails to go into alarm, quickly touch the same wire to the negative side of the standby battery. In either case, the faulted zone should go into alarm.

Hunt says that, if this procedure fails to activate an alarm condition, the problem may be in one of the bell circuits. To find out which one, remove one circuit at a time and replace it with the appropriate end-of-line resistor. Once the offending circuit has been isolated, Hunt says to use an ohm/volt meter to find the problem.

PROPERLY TEST YOUR SMOKE DETECTORS

Question: What is the proper procedure and test intervals for testing smoke detectors in commercial buildings? Also, is it all right to use a canned smoke spray to test smoke detectors?

Answer: NFPA 72, 1993 Edition, published by the National Fire Protection Association (NFPA), Quincy, MA, says that automatic smoke detectors should be visually inspected on a semi-annual basis and functionally tested once a year. Of course, when you test a smoke detector, you visually inspect it also, so a visual-only inspection must be performed within six months of a functional test.

Dealers also are supposed to conduct a sensitivity test on their smoke detectors within one year of installation and every other year after that. After the second year, if smoke detector sensitivity has not changed, the dealer can readjust the frequency of the calibration tests to every five years.

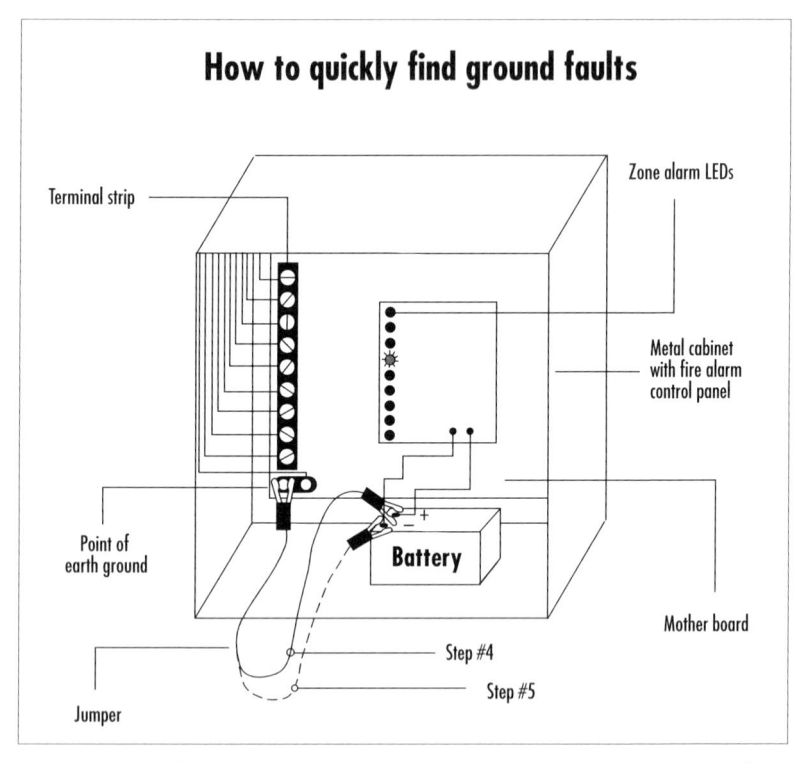

FIGURE 9–2. Allan C.W. Hunt, Installation Supervisor, Sentry Alarms Inc., illustrates how to isolate an earth-ground-faulted initiating circuit.

Servicing Fire Alarms 133

The NFPA is very specific in how dealers are supposed to test their smoke detectors. Always use the manufacturer's approved method. If the manufacturer of the smoke detectors that you install specifies an aerosol-type smoke product, then you can use it. However, in many cases, the manufacturer has a test tool that will functionally check the sensitivity of its smoke detectors. Before using a canned smoke spray, be sure that you read the instructions that came with the smoke detector.

In addition, before using an aerosol-type smoke product, be sure to read the instructions on the product itself so you use it correctly. Most manufacturers advise that at least 3 ft be maintained between the spray device and the smoke detector to assure that the chemical particles are dry before they enter the smoke chamber of a detector. This is important because wet particles can stick to the internal parts of the chamber, attracting dust and dirt. This can cause false alarms or even no alarm at all when there's a fire.

Then, after testing one or more smoke detectors, dealers should clean them to assure the removal of excess smoke chemical and dust particles. In some cases this is as simple as disassembling the detector and spraying the smoke chamber with a pressurized can of clean air. Canned air can be purchased at any commercial, wholesale electronics parts store.

EASY TEST FOR SMOKE DETECTORS

Open-area smoke detectors can be difficult to test when they are installed on high ceilings. Tim Wright, quality assurance manager for Alert Alarm Inc., Hilo, HW, says, "To test high-mounted smoke detectors without a ladder, tape a magnet to the tip of a 25-ft tape measure" (see Fig. 9–3, upper).

This method works well with smoke detectors that are equipped with a magnetic reed test-switch. "Raise the tape measure so it lines up with the smoke detector's test (reed) switch," says Wright. Holding the test magnet in place eventually will activate the detector.

Some smoke detectors have spring-loaded, momentary push buttons for a test switch. To test these detectors, "tape a $\frac{1}{8}$-in. drill bit to

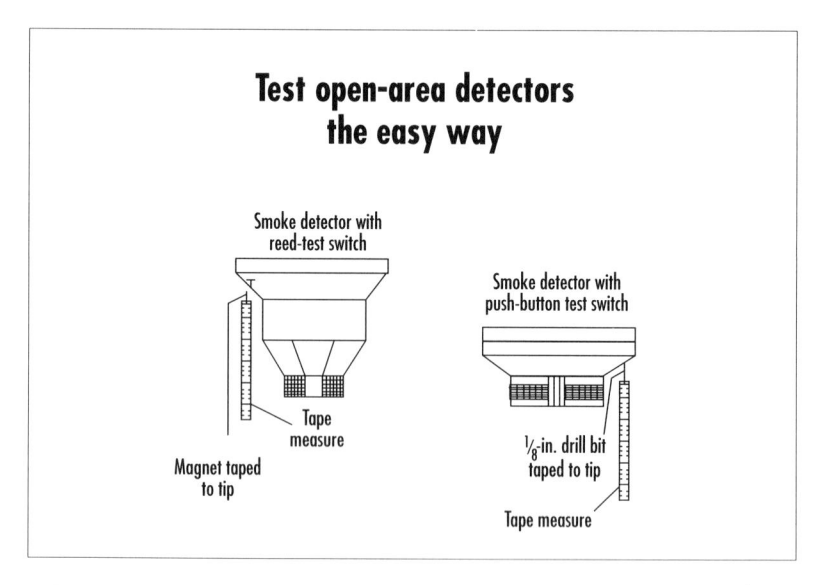

```
          Test open-area detectors
                the easy way

           Smoke detector with
           reed-test switch
                                    Smoke detector with
                                    push-button test switch


                    Tape
                    measure          ⅛-in. drill bit
         Magnet taped                taped to tip
         to tip
                                    Tape measure
```

FIGURE 9–3. Test smoke detectors the easy way by taping a magnet (left) or $\frac{1}{8}$-in. drill bit (right) to a tape measure.

the end of a tape measure. Be sure to stick the drill bit into the small hole at the tip of the tape measure" (see Fig. 9–3, lower).

If your tape measure will not support the weight of the magnet or drill bit over long distances (high ceilings), substitute a piece of $\frac{1}{2}$-in. galvanized, thin-wall conduit; a bell-hanger bit; or a long flexible drill bit for the tape measure.

FIND GROUND FAULTS USING THE HALF-CIRCUIT RULE

Use the half-circuit rule to troubleshoot ground faults, as well as shorts and bad connections, in long fire alarm circuits, says Bob Ruyle, president of Ruyle and Associates, Lincoln, NE. This service method enables alarm technicians to reduce guesswork, frustration, and troubleshooting time by isolating their circuit problem to a smaller portion of a circuit.

When servicing a fire alarm circuit experiencing intermittent ground-fault problems in a large facility, for example, begin by disconnecting the troublesome circuit from the initiating zone terminals in the fire alarm control panel. Then locate the smoke, heat, fire pull, or other detector that is physically halfway along the circuit. Start from this point in the circuit to make the job of troubleshooting more manageable (see Fig. 9–4).

After the halfway point in the circuit has been located, remove the smoke detectors (or smoke detector heads), heat detectors, and other detectors from the circuit. Then use jumper wires with alligator clips on each end to complete each leg of the circuit. This assures continuity from the beginning of the circuit at the fire alarm control panel to the end of the first circuit half and from the beginning of the second circuit half to the end-of-line device at the end of the troublesome circuit.

Clip one side of a sensitive ohm meter to the grounded leg and the other one to earth-ground and then work your way back toward the meter, disconnecting one jumper at a time until the problem suddenly disappears or the ohmage reading changes significantly.

Ruyle says that sometimes ground faults and shorts are difficult to find when they are in metal conduit. According to Ruyle, moisture that develops due to condensation eventually will find its way through defects in a conductor's sheath. A defect can occur during installation, for example, when installers pull the wires over a rough metal edge inside a conduit connector or coupling or the edge of the conduit itself. In this case, being at the right place at the right time is of utmost importance.

Then, after the offending length of wire or conduit is found, replace the wire altogether to permanently eliminate the ground fault.

INCREASING BATTERY POWER

Increasing battery voltage and amp-hour ratings is essential when expanding battery backup systems. Fire alarm systems quite often operate on 24 V, requiring that two 12-V or four 6-V batteries be connected in series, for example. Battery polarity must alternate so the voltage of each battery adds to the next, equaling the sum of all the batteries together (see Fig. 9–5).

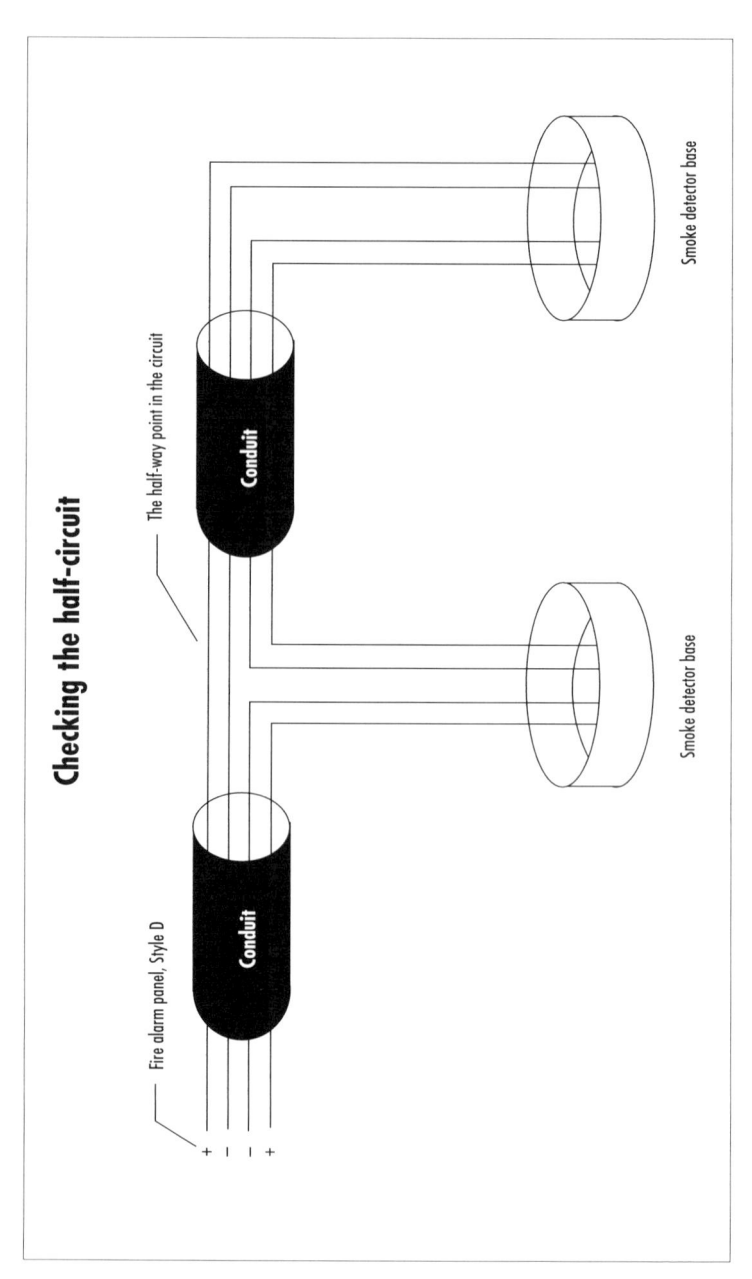

FIGURE 9-4. Divide long detection circuits in half to make the job of troubleshooting easier.

Servicing Fire Alarms

Multiply your battery power

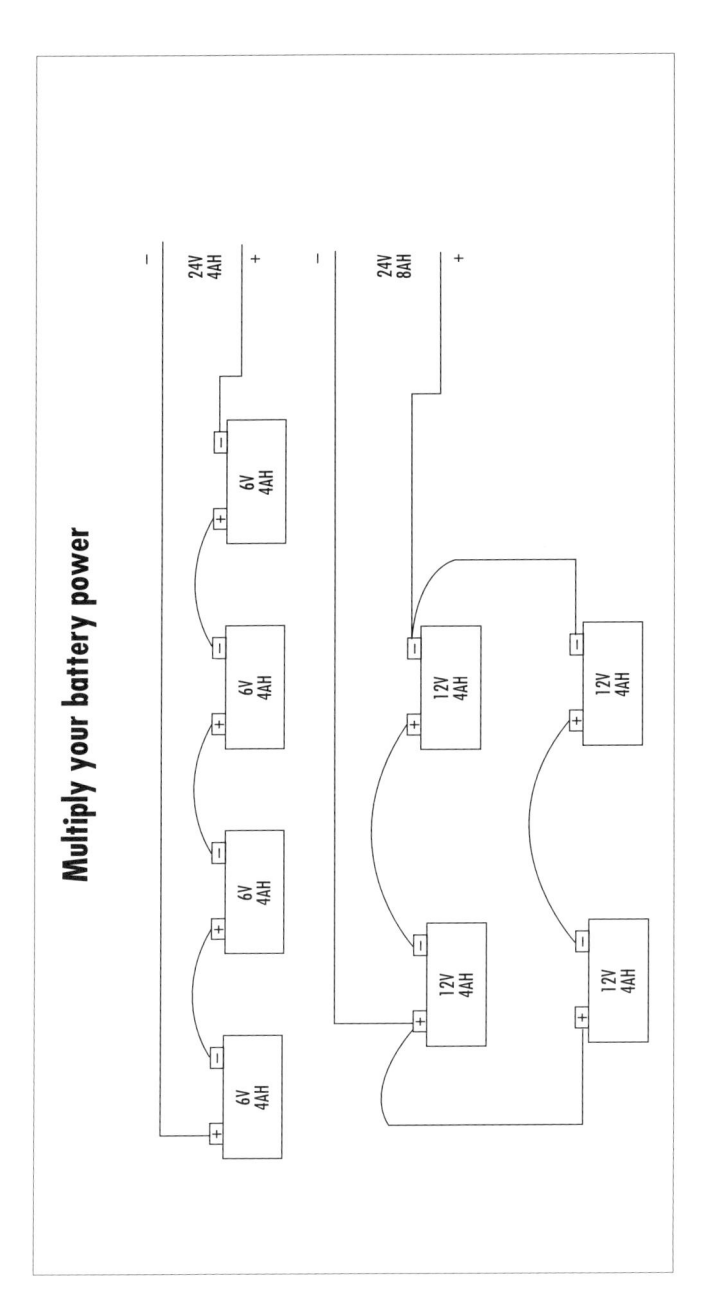

FIGURE 9-5. To increase the voltage of your battery supply, connect the batteries in series. To increase the power potential of your battery supply, connect the batteries in parallel.

The overall amp-hour rating of a battery backup system is increased by placing identical batteries in parallel. The amp-hour rating now increases while the voltage remains the same.

Both the amp-hour rating and the battery voltage are increased by placing a number of batteries in series, increasing the overall voltage, and using an identical series-connected set of batteries in parallel, doubling the overall amp-hour rating of the battery supply.

Index

A
AACC Security Systems, 25
Access control, 1
　installation, 3–7
　servicing, 9–19
Ac-induced false alarms,
　49–50
ACME Central Station Alarm Co., 46
Alarm and Stone Technologies Corp., 27
Alarmax Distributors Inc., 50
Alarm
　control panels, 23–33
　integrity, 31–32
　sensor tester, 82–84
Alarm Services, 26
Alert Alarm Inc., 133
Altronix Corp., 29
American Guard Dog, 42–43
Americans with Disabilities Act (ADA),
　106–107
　"adjoining room" clarification,
　107–108
　compliance with, 108–111
　PLC technology, 125–127

Antennas
　alignment of, 62–63
　zone wires as, 52
API Alarm Inc., 65
Arm-only capability, 45–46
Audible devices, 109–110
　door strike, 6–7
　outside, 43–44
Authority having jurisdiction (AHJ),
　3, 4, 100, 102, 112, 125
Automatic Fire Alarm Association Inc.
　(AFAA), 110, 113, 118

B
Back electromotive force (BEMF), 71
Battery power
　increasing, 17, 135, 137–138
　loss of, 44–45
　PBX systems, 42
Bell noise, 70–71
Block, Roger, 88, 89
Blueprints, filing of, 125
BOCA (Building Officials and Code
　Administrators Intl., Inc.), 108, 112

Burglar alarm installation, 43–45
    alarm-controlled guard dogs, 42–43
    EOLs, 23–29
    in garages, 34–39
    hold-up systems, 29–31
    magnetic surface switches, 34
    panel integrity, 31–32
    PBX systems, 42
    photobeam wires, 39–40
    switch nomenclature, 34
    tinning stranded wires, 32–33
    wireless transmitters, 40–41, 61–63
    zoning of, 46–47
Burglar alarms, 21
    preventing false alarms, 49–61
    servicing, 61–63
Burle Industries, Inc., 90

C

Cameras
    lightning protection, 88–90
    quad jitters, 88
    resolution, 87–88
    tube, 93–94
Capacitors, 71–73
Carson, Wayne, 126
Central Signaling Monitoring Co.
    (Cen Signal), 39
Central station monitor, 65–68
Cerberus Pyrotronics, 114
Certified Alarm Co., 43
Circuits
    double-break, 4–6
    feed and return, 102
Circuits, do-it-yourself
    alarm sensor tester, 82–84
    bell noise malfunction, 70–71
    capacitors, 71–73
    equipment load tester, 82
    MOSFET circuit, 80–81
    reed switches, 69–70
    siren-to-relay circuit, 81–82
    speakers and drivers, 77–78
    switch plates, use of, 73–75
    transistor relay, 78–80
    voltage regulators, 75–77

CleanTeam Co., 11
Closed circuit television (CCTV), 85
    coaxial cables, spacing of, 90–92
    ground loops, 92–93
    grounds, 94–95
    lightning protection, camera, 88–90
    quad system jitters, 88
    resolution, 87–88
    tube cameras, 93–94
    video monitors, setting up, 90
Coaxial cables
    grounding, 94–95
    installation of, 90–92
Code compliance, 112–113
Communication standard, SIA, 120
Condition codes, 66–67
Conductors, 124–125
Continuous-duty devices, 6
Cook, Bob, 5
Corporate Protection Services Inc., 3
Curwin, Bob, 87

D

Detection Systems Inc., 114
Dial-out delay, 65
Door strike releases
    phantom, 9
    quiet, 6–7
Double-break circuit, 4–6

E

Edwards, Div. General Signal, 114
Egress
    motion detectors, 12–14
    from public buildings, 3–6
Electromagnetic interference (EMI), 50–51
Electromagnetic locks, 4
End-of-line (EOL) resistors, 23–29, 51
    markers for, 100
Equipment load tester, 82

F

Fair-Rite Products Corp., 50, 51
False alarms, 44–45
    ac-induced, 49–50
    EMI problems, 50–51
    foil-tape, 57–59

Index

intermittent, 55, 57
microwave, 59
protection from, 59, 61
RFI problems, 51–55
False dispatch problems, 65–67
"Family," definition of, 99–100
Feed and return circuits, 102
Fiber-optic cable, 120
Fire alarm, 97
   ground fault problems, 131
   non-integration with access control, 3–4
   servicing, 129–138
   updating old, 116–117, 129–130
Fire alarm installation
   ADA and PLC technology, 125–127
   ADA requirements, 107–111
   beam-type smoke detectors, 113–115
   bedroom smoke detectors, 102–104
   blueprint filing, 125
   code compliance, 112–113
   conductor use, 124–125
   duct-type smoke detectors, 115–116
   electric smoke detector back-up, 18–120
   EOL markers, 100
   feed and return circuits, 102
   fiber-optic cabling, 120
   fire certificate, 100–102
   fire panel power, 122
   NFPA "family" definition, 99–100
   NICET certification, 111–112
   plug-in transformers, 127–128
   power outages, 120
   SIA communication standard, 120
   speaker installation, 122–124
   strobe intensity, 104–107
   supervisory signals, 121–122
   termination of wires, 121
   upgrading old fire pulls, 116–117
Fire panel power, 122
Fire pulls, upgrading, 116–117
First Inertia Switch Ltd., 114
Focus Publishing Enterprises, 120
Foil-tape false alarms, 57–59
Fuses, bad, 17–19

G
Gamewell Co., 114
Garage door openers, 38–39
Garages, protection of, 34–39
Ground faults, fire alarm, 131, 134–135
Ground loops, 92–93, 95
Grounds, 89–90, 94–95
Guard dogs, 42–43
Guide to Code Requirements, A, 113
Gyyr Service, 88

H
High-voltage wires, precautions regarding, 90–92
Hochiki America Corp., 114
Hold-up systems, 29–31
Home Services Inc., 55
Hunt, Allan C.W., 131
HVAC systems, 116

I
ICBO (International Conference of Building Officials), 112
Image Marketing, 87
Impedance-matching transformers, 124
Independent Alarms & Monitoring, 59
Interactive Technologies Inc. (ITI), 41

K
Keyswitches, use of, 73–75

L
Larman, Wally, 42
LEDs, use of, 74–75
Leviton Telcom, 32
Life Safety Code, 1, 12, 14, 99, 112, 113
Lightning protection, 94–95
   camera, 88–90
LRC Electronics Co., 93

M
Macro Media Systems Inc., 127
Madden, Ken, 88
Magnetic-stripe card
   care of, 10–11
   misreads, 10
   reader clean-up, 11

Magnetic surface switches, 34
Markowitz, Nick, 15, 16, 17, 26, 29, 36, 45, 46, 73–74, 122, 129–130
Markowitz Electric Protection, 15, 16, 26, 29, 36, 45, 73, 122, 129
Mayer Engineering Associates Inc., 108
Mayer, Kenneth, 108
Meter-reading device, automatic, 63
Microwave false alarms, 59
Minieri, Mike, 105
Mioduszewski, Frank, 114–115
Moore-Wilson Signaling Report, The, 117, 120
MOSFET (metal-oxide semiconductor field effect transistor) circuit, 80–81
Motion detectors
    egress, 12–14
    microwave, 59, 60
Multiple-zone security system, 46–47
Mutual Central Station Alarm Services Inc., 42

N

Nascom Inc., 69
National Electrical Manufacturers Association (NEMA), 34, 113
National Electric Code, 17, 112, 122, 124
National Fire Alarm Code, 1, 100, 102, 104, 112, 115, 127
National Fire Protection Association (NFPA), 1, 14, 104, 112, 115, 116, 118, 122, 125, 127, 132, 133
    "family" definition, 99–100
National Institute for Certification in Engineering (NICET), 111–112
NC detection devices, 51
Neibauer, Larry, 110, 113, 118
NTSC (National Television Standards Committee), 87

O

Office on the Americans with Disabilities Act, 111
Optex International Group, 114

P

Passive infrared detectors (PIRs), 12–13, 34, 36–37, 59, 61
PBX systems, 42
Photobeam wires, 39–40
PhotoScan Northwest, 88
Pixel-to-TVL formula, 88
PLC technology, 125–127
PolyPhaser Corp., 88–89, 95
Power outages, 45, 120
Professional Alarm Service Organizations of North America Inc. (PASONA), 101–102, 121, 127
Programming error, 65–67
Pro-Tech Systems Inc., 116
Public
    accommodation, 110
    buildings, egress from, 3–6

Q

Q-Card Inc., 10, 11
Quad system jitters, 88
Quality Control of Automatic Fire Detection and Alarm System Installations, 113

R

Radio Amateur's Handbook, The, 16, 17
Radio-frequency interference (RFI), 51–55
    sources of, 56
Radio Shack, 76, 82, 84
Relay sparking, halting, 14–15
Remote switch, 6
Renwick, Jim, 95
Resolution, 87–88
RJ31X jacks, 63, 67–68
Ruyle, Bob, 134
Ruyle and Associates, 134

S

SBCCI (Southern Building Code Congress Intl.), 112
Securitron Magnalock Corp., 5
Security Industry Association (SIA), 120
Sentry Alarms Inc., 131

Index

Shielding, cable, 92
Single-Pole-Double-Throw (SPDT) switches, 5–6, 26, 71
Siren-to-relay circuit, 81–82
Smoke detectors
    backing up electric, 118–120
    beam-type, 113–115
    bedroom, 102–104
    duct-type, 115–116
    testing, 131–134
Sparkless relay, 15
Speakers
    fire alarm connection, 122–124
    matching drivers with, 77–78
Stair towers, pressurization of, 3
Stranded wires, tinning of, 32–33
Strobe intensity, calculation of, 104–107
Supervisory signals, 121–122
Surge protectors, 88–89
Switches
    defective, 55, 57
    nomenclature, 34
    reed, 69–70
System Sensor, 114

T

Taylor, Harvey, 88
Telephone line, routing of, 67–68
TOFLAN CORP., 105
Transformers
    calculating size, 14
    impedance-matching, 124
    plug-in, 127–128
    ventilation of enclosures, 15–17
Transistor relay circuit, 78–80
Transmitters, wireless, 40–41, 61–63
Tube cameras, 93–94
Twisted-pair cable, 31–32

U

Underwriters Laboratories Inc., 101, 104, 105, 106

V

Video monitors, set up of, 90
Visible devices, 109–110
Voltage
    regulation, 75–77
    spike, 71, 88
Vydra, Jim, 116

W

Walkie-talkies, testing with, 55, 57
Wattage, speaker, 78
Williams, Jim, 55
Wireless transmitters, 40–41, 61–63
Wire termination, 121
Wright, Tim, 133

X, Y, Z

Yamaha Corp. of America, 93
Zoning, alarm, 27–29, 46–47

Other Books from Butterworth-Heinemann

Effective Physical Security, Second Edition
Lawrence J. Fennelly
1996   256pp   pb   0-7506-9873-X

Executive Protection Professional's Manual, The
Philip T. Holder and Donna Lea Hawley
1997   168pp   pb   0-7506-9868-3

Formula for Selling Alarm Systems, The
Lou Sepulveda
1996   104pp   pb   0-7506-9752-0

Home Security, Second Edition
Vivian Capel
1997   192pp   pb   0-7506-3546-0

Security, ID Systems and Locks: The Book on Electronic Access Control
Joel Konicek and Karen Little
1997   245pp   pb   0-7506-9932-9

Detailed information on these and all other BH-Security titles may be found in the BH-Security catalog(Item #800). To request a copy, call 1-800-366-2665. You can also visit our web site at: http://www.bh.com

These books are available from all good bookstores or in case of difficulty call: 1-800-366-2665 in the U.S. or +44-1865-310366 in Europe.

E-Mail Mailing List
An e-mail mailing list giving information on latest releases, special promotions/ offers and other news relating to BH-Security titles is available. To subscribe, send an e-mail message to majordomo@world.std.com. Include in message body (not in subject line) subscribe bh-security